PLANNING UNDER UNCERTAINTY

Solving Large-Scale Stochastic Linear Programs

GERD INFANGER
Stanford University

boyd & fraser publishing company

Senior Acquisitions Editor: DeVilla Williams
Production Services: Publications Services
Compositor: Publications Services
Interior Design: Publications Services
Cover Design: Michael Rogondino
Manufacturing Coordinator: Tracy Megison

©1994 by Gerd Infanger
△ The Scientific Press Series
A Division of South-Western Publishing Co.
One Corporate Place ● Ferncroft Village
Danvers, Massachusetts 01923

International Thomson Publishing
boyd & fraser publishing company is an ITP company.
The ITP trademark is used under license.

Manufactured in the United States of America.

Library of Congress Cataloging-in-Publication Data

Infanger, Gerd.
    Planning under uncertainty : solving large-scale stochastic linear
programs / Gerd Infanger.
      p. cm. — (The Scientific Press series)
    Includes bibliographical references (p. –)
    ISBN 0-89426-249-1
    1. Stochastic programming. 2. Linear programming. I. Title.
II. Series
T57.79.I53 1993
658.4′033–dc20
                                    93-46658
                                        CIP

1 2 3 4 5 6 7 8 9 10 D 7 6 5 4 3

To my wife Massy
and my parents
Josef and Maria Infanger

Foreword

Optimization under uncertainty represents an important advance in man's ability to better plan and at the same time hedge against the uncertain bad events that might happen in the future. This is a fundamental problem of decision science, and this book is all about ways to formulate and solve it.

It is a pleasure for me, as a pioneer in the field of programming under uncertainty, to say a few words about the author. I first met Gerd Infanger in 1989, when he came to Stanford as a visiting scholar from the Technical University of Vienna. He and I have been collaborating very closely ever since. He leads my little group in the Operations Research Department's System Optimization Laboratory, where we explore the new methodology in this field.

Gerd Infanger is now one of the leading experts in the world developing large-scale optimization models for planning under uncertainty. Since 1985 this field has grown to be very popular, with many contributors all over the world. In this book he reviews their work as well as his own; he discusses how to formulate and apply the models and presents the algorithms that he and the others have developed to solve them.

From the very beginning of his research, Infanger became intrigued with the idea of combining the decomposition principle with importance sampling to solve large-scale linear problems whose coefficients and right-hand sides are not known with certainty. He assembled the analytic tools from mathematics, statistics, and computer science that were needed to implement the approach, and he tested the software on practical problems. We expected, because of the enormous size of the practical problems (when expressed as equivalent deterministic linear programs), that it would be possible to solve them only on mainframes or parallel processors, but it turned out that the importance sampling was so powerful that it made solving them on personal computers practical. In this book Infanger reports on his success in formulating and solving these practical problems.

This discovery, that it is practical to solve linear problems under uncertainty on PCs, has opened the door to a world of new applications that will eventually change the way planning is done.

George B. Dantzig
Stanford University
September 15, 1993

Contents

Preface

An important class of problems concerning optimal allocation of scarce resources over time can be formulated as dynamic linear systems. Although efficient techniques have been developed to solve large deterministic dynamic linear systems, the solutions obtained from these systems have turned out to be unsatisfactory because they failed to hedge against certain contingencies that could occur in the future. Stochastic models address these shortcomings but have not been widely used so far because they seemed intractable because of their size.

A novel approach originated by G. B. Dantzig and P. Glynn and developed jointly by them and the author combines classical decomposition techniques with a relatively new technique called importance sampling. Our approach has turned out to be capable of solving large-scale stochastic linear programs with numerous stochastic parameters which have hitherto been unsolvable.

The basic concept for solving two-stage stochastic linear programs is as follows. Using dual (Benders) decomposition, we decompose the stochastic linear program into a master problem (the first-stage problem) and a series of subproblems (the second-stage problems). Each subproblem corresponds to a scenario—that is, a certain combination of outcomes of the uncertain parameters in the model.

We iteratively solve the master problem to obtain a first-stage trial solution for the problem and a series of subproblems to evaluate that trial solution, that is, to obtain the expected second-stage costs and their sensitivity with respect to the trial solution. The second-stage information is passed to the master problem in the form of Benders cuts; an updated trial solution is obtained in each Benders iteration. The algorithm terminates when a particular trial solution can be declared optimal. Instead of solving all possible subproblems to compute the expected future costs exactly (which is impossible even for a small number of uncertain parameters), we use Monte Carlo importance sampling techniques to obtain estimates of the expected future costs and their sensitivity. Importance sampling is crucial in our concept. It is a powerful variance reduction technique used to obtain accurate estimates with only a small sample size. Thus, the number of subproblems that has to be solved in each Benders iteration can be kept small.

We first developed the theory for solving general two-stage stochastic linear programs with recourse and a certain restricted class of multi-stage problems, for which the problem breaks down into two parts: a deterministic dynamic part and a stochastic part. The latter structure arises from the formulation of facility expansion-planning problems. Our implementation uses MINOS as a subroutine for solving linear subproblems. Numerical results from large-scale test problems in the areas of facility expansion planning and financial planning demonstrated that very accurate solutions of stochastic linear programs can be obtained with only a small sample size.

The large-scale test problems included various stochastic parameters. For example, the largest problem representing expansion planning for multiarea electric power systems included 39 stochastic parameters. In the deterministic equivalent formulation, if it were possible to state it, the problem would appear as a linear program with about 4.5 billion constraints and variables. The largest portfolio optimization problem included 52 stochastic parameters, which in the deterministic equivalent formulation would appear as a linear program with about 10^{27} constraints and a similar number of variables. Problems this large hitherto seemed to be intractable. Using our method, however, we have been able to solve them on a laptop 80386 computer.

The test results indicate that we have not yet reached the limits of the approach. Necessary sample sizes turned out to be so small that use of parallel processors proves not to be a *sine qua non* condition for solving even large-scale stochastic linear problems. In order to speed up the computation time in the case where large sample sizes are required, we have developed a parallel implementation running on a hypercube multicomputer. The numerical results show that speedups of about 60% can be obtained using 64 parallel processors.

Encouraged by the promising numerical results for two-stage and a restricted class of multi-stage problems, we have developed the theory for a general class of multi-stage stochastic linear programs. Our approach for solving multi-stage problems includes special sampling techniques for computing upper bounds and methods of sharing cuts between different subproblems. It will enable us to efficiently solve large-scale problems with many stages and numerous stochastic parameters in each stage. The implementation is subject to future research, but preliminary numerical results have turned out to be promising.

Further research includes improved decomposition techniques for large-scale problems (e.g., optimized tree-traversing strategies and passing information based on nonoptimal subproblems), improvements to the importance sampling approach (e.g., using different types of approximation functions), improved software (e.g., a parallel implementation of the multi-stage algorithm on distributed workstations), and the testing of the methodology on different practical problems in different areas.

Acknowledgments

Planning under uncertainty and solving stochastic problems by combining decomposition techniques, Monte Carlo sampling techniques and parallel processors is a theme composed by Professor George B. Dantzig, and the use of importance sampling instead of crude Monte Carlo sampling has been recommended by Professor Peter W. Glynn. At the beginning of 1989, the author joined the Department of Operations Research at Stanford University as a visiting scholar from Vienna University of Technology and started to work closely with Professor Dantzig on the development of techniques for solving large-scale stochastic linear programs. Progress on this topic has led to extensions of the author's visit and to his current position of Senior Research Associate at the Department of Operations Research at Stanford University.

It has been an outstanding and rewarding experience for the author to know and collaborate with Professor George Dantzig. The author wants to thank Professor Dantzig for everything—all his advice, his outstanding professional support, and his great friendship.

The author wants to thank Professor Peter Glynn for the many valuable discussions on the topic of planning under uncertainty and importance sampling and his great support.

The author is grateful to Professor Michael Saunders and Dr. John Stone for their valuable discussions on the topic of large-scale systems and their important and helpful suggestions concerning previous versions of this work.

The author collaborated with Professor James K. Ho on using parallel processors. The author wishes to thank Professor Ho for this collaboration.

Thanks also to Dr. David Morton for valuable comments on a previous version of this work and to Dr. Alamuru Krishna, who assisted in preparing some of the test problems.

The author is especially grateful to Professor Peter Jansen, his "Doktorvater," for initiating the author's visit to Stanford University and his continuing great support.

Professors George B. Dantzig, Peter Jansen, John Mulvey, and Dr. John Stone served as reviewers of the monograph. The author wishes to thank them for their effort. The author is grateful to Peter Fairchild, Lorraine Metcalf, Devilla Williams, and others for making the production of this book possible.

1

Introduction

1.1 DYNAMIC SYSTEMS UNDER UNCERTAINTY

A fundamental economic problem is the optimal allocation of scarce resources over time. Since Dantzig (1948) [20] invented the simplex method for linear programming (see Dantzig (1963) [22]), Operations Research has been developing efficient techniques to address this important problem. Important developments of Operations Research include recent advances in linear, nonlinear, and discrete optimization techniques, and especially the advancements of large-scale optimization techniques. For basic references, see Gill, Murray, and Wright (1981) [56], (1991) [57], Lasdon (1970) [86], and Geoffrion (1974) [55]. Recent advances in computer technology, e.g., vector processing and distributed computation on parallel processors, further contribute to the capabilities of modern Operations Research methods.

An important class of resource allocation problems over time can be formulated as dynamic linear systems. These are linear programs with a certain matrix structure. The nonzero elements of the constraint matrix appear in a staircase pattern, with each step of the staircase corresponding to a certain time period.

This structure emerges because constraints associated with a particular time period have coefficients in that time period and in the period before, for all time periods of the planning horizon. Systems of this kind are also called multi-stage linear programs. Note that optimal linear control problems in discrete time also fall into this category.

Multi-stage linear programs have been studied extensively. Methods for solving staircase systems exploit their structure in order to increase computational efficiency. Dantzig and Wolfe's (1960) [31] primal decomposition and nested dual decomposition based on Benders (1962) [6] have been exploited and further developed. Ho and Loute (1981) [69] gave efficient implementations and collected a set of staircase linear programming test problems. Some references for techniques for solving dynamic systems are Glassey (1973) [65], Ho and Manne (1974) [70], Abrahamson (1983) [1], and Wittrock (1983) [130]. Parallel dual decomposition has been applied by Entriken (1988) [39], (1989) [40], and parallel primal decomposition by Ho and Gnanendran (1989) [67] and Ho, Lee, and Sundarraj (1988)[68].

Although efficient techniques allow us to solve dynamic systems of very large sizes (e.g., problems can be solved with several hundred thousand variables), the solutions obtained from these systems have often proved impractical. Parameters of the system, which had been assumed to have certain values when the model was formulated, assumed different values when the optimal solution was finally implemented. These were viewed as deterministic modeling systems, where all parameters were assumed to be known to the planner with certainty. Solutions obtained from deterministic planning models fail to hedge against different contingencies that occur unpredictably in the future.

Consider, for example, an operations-planning problem of an electric power system. An operations plan is determined, subject to constraints based on certain availabilities of generators and transmission lines and on certain values of demands. The optimal solution obtained is optimal only for this particular choice of parameters. When the optimal solution is implemented, the generators and transmission lines may assume different values of availability (e.g., due to unplanned failures), and demands may result that are different from the planned values. For these realized values of the parameters of the system, the solution is no longer optimal. Costs different from those planned are the consequence. In extreme situations the implemented solution may lead to infeasibility; for example, the demands may be unsatisfiable, and very costly recourse actions may be necessary.

As a further example, consider the portfolio optimization problem. A deterministic approach would assume the returns of the equities traded at the market as known parameters. The optimal solution of the deterministic model would contain fractions of equities in decreasing order of their planned returns, subject to the constraints and bounds in the model. It would contain as many shares as possible of the equity with the highest assumed return, then as many as possible with the second highest return, and so forth. An investor implementing this optimal solution (that is, having bought the recommended numbers of shares at the market) may be disappointed when he later observes different values of returns than have been assumed in the model. The returns of a portfolio selected by a deterministic model may be significantly different from those expected. Of course, no one would attempt to implement the solution of a deterministic portfolio optimization problem.

Because solutions obtained from deterministic planning models have turned out to be unsatisfactory, different techniques have been developed to compensate for their shortcomings.

- Sensitivity analysis examines the changes of the optimal solution and the optimal objective value with respect to variations of uncertain parameters that are considered to be important. Sensitivity analysis is usually conducted by varying one parameter at a time. Linear programming theory provides for local sensitivity results as a postoptimal analysis; see Dantzig (1963) [22]. Usually, not only is local sensitivity analysis performed, but uncertain parameters are varied according to their full range of possible outcomes. If the solution of the problem turns out to be very sensitive to a particular parameter, the value of this parameter is revised to be more on the safe side. Then, by solving the optimization problem again, one hopes to find a new solution that accounts, at least partially, for the uncertainty. Clearly, sensitivity analysis cannot fully overcome the shortcomings of deterministic planning models.

- Scenario analysis is another method that has been widely used and seems to be the preferred technique for many planners. In this approach one assumes scenarios (certain combinations of possible values of the uncertain parameters) and solves the problem for each. The different optimal solutions and the corresponding optimal objective values in the different scenarios are then aggregated in a heuristic way. By conducting scenario analysis, the planner hopes to get insight into the problem. By solving the problem repeatedly for different scenarios and studying the solutions obtained, the planner observes sensitivities and heuristically decides on an appropriate solution.

- Worst-case analysis and other related techniques attempt to account for uncertainty by putting safety margins into the problem formulation.

On the other hand, stochastic models address the shortcomings of deterministic models directly. Instead of assuming the uncertain parameters of the model to be known, stochastic models assume their *distributions* to be known. This causes the model size to grow enormously, as will be discussed in detail later. As a result, it has been too difficult to solve real-world stochastic models, and they have not yet been widely used. A new approach, based on decomposition techniques and Monte Carlo importance sampling, has been created by Dantzig and Glynn (1990) [24] and developed by them and Infanger (1992) [74]. We believe that this approach is a breakthrough in solving problems of planning under uncertainty. In the following we will discuss this approach.

We first outline the basic theory for two-stage stochastic linear problems and extend the discussion to a special class of multi-stage problems. We demonstrate the power of the method with numerical results obtained from different test problems. We then discuss an implementation of the method, using parallel processors, that we have developed in collaboration with Ho (Dantzig, Ho, and Infanger (1991) [26]). Finally, we derive the theory of solving general multi-stage stochastic linear problems.

We begin by introducing the stochastic problem and discussing different approaches to its solution.

1.2   THE STOCHASTIC OPTIMIZATION PROBLEM

Defining a probability space (Ω, P), where Ω is the set of possible realizations ω of the uncertain parameters and P the corresponding probability distribution, a stochastic optimization problem can be represented as

$$\begin{aligned} z \;=\; & \min\; E(f(x,\omega)) = \int f(x,\omega)\,dP(\omega) \\ & \text{s/t}\; x \in C \subseteq R^n, \end{aligned} \tag{1.1}$$

where $f(x, \omega)$ is the objective function and C the set of feasible solutions defined by the constraints of the optimization problem. The probability space (Ω, P) can represent all kinds of distributions, e.g., continuous or discrete, with a finite or infinite number of realizations. For example, consider the case in which Ω is discrete and finite and a discrete outcome ω has corresponding probability $p(\omega)$. Problem (1.1) then is represented as

$$
\begin{aligned}
z \quad = \quad & \min \ E \ f(x, \omega) = \sum_{\omega \in \Omega} f(x, \omega) p(\omega) \\
& \text{s/t } x \in C \subseteq R^n.
\end{aligned}
\tag{1.2}
$$

Problems (1.1) and (1.2) are hard to solve. The main problem lies in the sheer number of possible outcomes that have to be taken into account. In problem (1.2) we define the number of elements in Ω to be $K = |\Omega|$. In practical problems, K is a very large number. In order to fully understand the nature of a stochastic solution, we will discuss different solution approaches to the stochastic optimization problem. For simplicity and ease of exposition we conduct the discussion for the case of discrete and finite Ω.

1.2.1 The "Wait and See" Approach

In the "wait and see" approach we assume that we can somehow wait until the uncertainty is resolved at the end of the planning horizon, and an outcome $\omega \in \Omega$ can be observed, before we make the optimal decision x. The "wait and see" approach therefore assumes perfect information about the future. It is clear that such a solution is not implementable. The corresponding problem can be stated as

$$
\begin{aligned}
z^\omega \quad = \quad & \min \ f(x, \omega) \\
& \text{s/t } x \in C^\omega \subseteq R^n,
\end{aligned}
\tag{1.3}
$$

$$
x^\omega \in \arg\min \ \{ f(x, \omega) \mid x \in C^\omega \},
\tag{1.4}
$$

$$
z_{ws} = E \ z^\omega = \sum_{\omega \in \Omega} z^\omega p(\omega).
\tag{1.5}
$$

The set of feasible solutions of x, C^ω, is defined by the constraints in scenario ω. We solve the problem for the observed outcome ω and obtain an optimal solution subject to scenario ω, x^ω. It is an optimal solution given perfect information about the future. We compute the expected value of the optimal costs z^ω to obtain z_{ws}, the expected costs under perfect information.

1.2.2   The "Here and Now" Approach

The "here and now" approach represents the true stochastic optimization problem of (1.2). A decision x has to be made "here and now," before observing an outcome from Ω. The value x is chosen such that the expected costs $E\ f(x,\omega)$ assume a minimum:

$$
\begin{aligned}
z \ &= \ \min\ E\ f(x,\omega) \\
&\quad \text{s/t } x \in C = \bigcap_{\omega \in \Omega} C^\omega.
\end{aligned}
\tag{1.6}
$$

The optimal objective function value z denotes the minimum expected costs of the stochastic optimization problem. Note that x has to be feasible for all scenarios $\omega \in \Omega$; thus $C = \bigcap_{\omega \in \Omega} C^\omega$ denotes the intersection of all C^ω, $\omega \in \Omega$, where C^ω represents the feasible region given by the constraints in scenario ω. The optimal solution

$$
x^* \in \arg\min \ \{E\ f(x,\omega) \mid x \in \bigcap_{\omega \in \Omega} C^\omega\}
\tag{1.7}
$$

represents the realistic solution of the stochastic optimization problem. The solution x^* hedges against all possible contingencies $\omega \in \Omega$ that may occur in the future.

1.2.3   The Expected-Value Approach

Let $\bar{\omega}$ denote the expectation over the set Ω:

$$
\bar{\omega} = E\ \omega = \sum_{\omega \in \Omega} \omega p(\omega).
\tag{1.8}
$$

In the expected-value approach we replace the stochastic parameters by their expected values and solve the corresponding deterministic problem:

$$\hat{z}_d = \begin{array}{l} \min\ f(x, \bar{\omega}) \\ \text{s/t } x \in C^{\bar{\omega}}, \end{array} \tag{1.9}$$

$$x_d \in \arg\min\ \{f(x, \bar{\omega}) \mid x \in C^{\bar{\omega}}\}. \tag{1.10}$$

We refer to problem (1.9) as the **expected-value problem** corresponding to the stochastic optimization problem (1.2). $\hat{z}_d$ denotes the costs corresponding to x_d, the optimal solution of the expected-value problem. We denote by z_d the expected costs corresponding to the implementation of the solution of the expected-value problem, x_d:

$$z_d = E\ f(x_d, \omega). \tag{1.11}$$

Note that $x_d \in C^{\bar{\omega}}$ does not necessarily imply that $x_d \in C^{\omega}$, $\omega \in \Omega$. We let $z_d \to \infty$ if $x_d \notin C^{\omega}$, $\omega \in \Omega$.

1.2.4 Assessment of the Different Approaches

Assuming $f(x, \omega)$ to be convex and comparing the expected costs of the different solutions of the stochastic optimization problem, we can show that

$$z_{ws} \leq z \leq z_d. \tag{1.12}$$

We define the expected value of perfect information, EVPI, to be

$$\text{EVPI} = z - z_{ws}. \tag{1.13}$$

It is a measure of how much one would be willing to pay (at most) to obtain perfect information about the future. A small EVPI indicates that refined forecasts will lead to little gain. A large EVPI indicates that incomplete information about the future is costly.

We define the value of the stochastic solution, VSS, to be

$$\text{VSS} = z_d - z. \tag{1.14}$$

It is a measure of how much can be saved by implementing the solution of the stochastic optimization problem versus the solution of the deterministic expected-value problem. If VSS is small, the approximation of the stochastic problem by the corresponding expected-value problem is a good one and the obtained expected-value solution is a good solution for the stochastic problem. The larger the value of VSS, the more important it is to obtain the solution of the stochastic optimization problem.

Birge (1980) [8] developed bounds on the expected value of perfect information and on the value of the stochastic solution; e.g.,

$$0 \leq \text{EVPI} \leq z - \hat{z}_d \leq z_d - \hat{z}_d, \tag{1.15}$$

$$0 \leq \text{VSS} \leq z_d - \hat{z}_d. \tag{1.16}$$

The bounds on EVPI and VSS are useful for deciding if it is necessary to solve the stochastic problem or if an approximation is adequate.

1.2.5   The Classical Stochastic Linear Program with Recourse

A classical stochastic optimization problem is the two-stage stochastic linear program with recourse. It has the form

$$
\begin{aligned}
z = \quad \min \quad & cx \; + \; E_\omega\, Q(x,\omega) \\
\text{s/t} \quad & Ax \qquad\qquad\qquad = \; b \\
& x \; \geq \; 0,
\end{aligned}
\tag{1.17}
$$

where

$$
\begin{aligned}
Q(x,\omega) \;=\; \min \quad & f(\omega)y \\
\text{s/t} \quad D(\omega)y \;=\;& d(\omega) + B(\omega)x \\
y \;\geq\;& 0.
\end{aligned}
\tag{1.18}
$$

The matrix A and the vector b are known with certainty. E_ω denotes the expectation with respect to ω, an element of the probability space (Ω, P), and the function $Q(x,\omega)$, referred to as the recourse function, is in turn defined by the linear program (1.18). The technology matrix $D(\omega)$, the right-hand side $d(\omega)$, the transition matrix $B(\omega)$, and the objective function coefficients $f(\omega)$ of this linear program are random. Since Dantzig (1955) [21], this type of problem has been studied extensively by many authors. A special class of (1.18) is that of complete recourse, where the recourse function $Q(x,\omega)$ is finite (i.e., (1.18) is feasible) for any choice of x. Properties of this class of problem have been studied by Wets (1966)[124]. The recourse function $Q(x,\omega)$ can be shown to be convex and is in general nonsmooth.

To evaluate the expectation $E_\omega Q(x,\omega)$, a multiple integral or a multiple sum typically has to be computed. We will show this later. The main difficulties in stochastic optimization deal with the evaluations of multiple integrals or multiple sums. The numerical computation of expectations requires large numbers of function evaluations, where each function evaluation means that a linear program has to be solved.

There are several approaches to this problem. In the following we give an overview of different approaches that have been developed to solve stochastic optimization problems.

1.2.6   Research in Stochastic Optimization

Stochastic programming with recourse was first introduced independently by Dantzig (1955) [21] and Beale (1955) [4]. Chance-constrained stochastic programming, involving a different type of model, was first introduced by Charnes and Cooper (1959) [18]. It involves models in which a decision is made prior to the knowledge of outcomes of random parameters, such that certain constraints are met with certain probability levels; see Prékopa (1988) [109] for an overview of probabilistic constrained programming models. A number of different algorithmic approaches have been proposed for solving two-stage stochastic linear programs of types (1.17) and (1.18), stated above. (See Kall (1976) [79], Wets (1974) [125], and Wets (1983) [126] for an investigation of the recourse problem.)

Van Slyke and Wets (1969) [122] showed with their L-shaped method how Benders (1962) [6] decomposition can be applied to solving two-stage stochastic linear programs. Their algorithm uses expected-value cuts, representing an outer linearization of the expected second-stage costs (or the recourse function). A variant proposed later by Birge and Louveaux (1985) [11] is based on multiple cuts, where a different cut with respect to each scenario is computed and the expected-value calculation is carried out in the master problem. Earlier, Dantzig and Madansky (1961) [29] pointed out that the dual of the two-stage stochastic linear program has a structure ideal for Dantzig-Wolfe (1960) [31] decomposition.

A classical approximation scheme for solving two-stage stochastic linear programs with stochastic right-hand sides (randomness in the transition matrix B and the right-hand side d; see equation (1.18)) is to calculate lower and upper bounds via the inequalities of Jensen (1906) [77], Edmundson (1956) [38], and Madansky (1959) [91], respectively, and to successively improve these bounds. (See Kall, Ruszczynski, and Frauendorfer (1988) [81].) The lower bound based on Jensen's inequality involves the evaluation of the recourse function at only one point, namely the expected value of the random parameters. The upper bound due to Edmundson and Madansky is based on the theory of moment spaces and is computed by weighting the extreme points of the support of the random variables. Refinements of this bound have been proposed by Ben Tal and Hochman (1972) [7], Kall (1974) [78], Huang, Ziemba, and Ben-Tal (1977) [71], Kall and Stoyan (1982) [82], and Frauendorfer and Kall (1988) [50] for independent random variables, and Dupačówá (1978) [36], Gassmann and Ziemba (1986) [54], Frauendorfer (1988) [48], and Birge and Wallace (1988) [13] for dependent random variables. Frauendorfer (1992) [49] uses Barycentric Approximations to solve two-stage stochastic programs. Ermoliev, Gaivoronski, and Nedeva (1985) [45] provided a general framework for stochastic programming problems. Birge and Wets (1987) [15] and Cipra (1985) [19] computed bounds based on solving a generalized moment problem. Based on that, Birge and Wets (1986) [14], (1989) [16] exploited the sublinear property of the recourse function. Wallace (1987) [123] proposed a procedure for the case that the evaluation of the recourse function involves the solution of a network problem, and Birge and Wallace (1988) [13] provided a separable piecewise linear upper bound. Upper bounds for the expectation of convex functions with discrete random variables and the relationship of moment problems and linear programming have been investigated by Prékopa (1988) [110], (1989) [111], (1990) [112]. Birge (1984) [9] proposed row and column aggregation schemes to approximate a stochastic program. Robinson and Wets (1987) [114] presented stability results for two-stage stochastic programs.

Using mathematical programming techniques seemed to be promising in special cases, e.g., Nazareth and Wets (1986) [100]. Strazicky (1980) [119] and Kall (1979) [80] proposed basis factorization approaches. Wets (1988) [128] surveyed the use of large-scale linear programming techniques, and Nazareth and Wets (1988) [101] provided an overview of using nonlinear programming techniques for solving stochastic programs. Ruszczynski (1986) [117] proposed his regularized decomposition method. Lustig et al. (1991) [90] empirically studied the performance of interior-point linear programming solvers on different formulations of two-stage stochastic linear programs.

Rockafellar and Wets (1989) [115] presented "Progressive Hedging," in which non-anticipativity constraints are enforced via Lagrangian penalty terms in the objective, and in which a two-stage or multi-stage program is solved for each scenario in each iteration. It provides an iterative method of solving scenario optimization problems to construct the solution of the stochastic problem. Numerical results of the performance of "Progressive Hedging" have been given by Mulvey and Vladimirou (1991) [97], [98].

Stochastic quasigradient methods select sequentially random search directions based on a limited number of observations of the random function (1.18) in each iteration. They have been studied by Ermoliev (1983) [42], (1988) [43] [44] and Gaivoronski (1988) [51]. The convergence rates of stochastic quasigradient methods are slow; it is important that objective values, subgradients, and stepsizes can be specified well. Pflug (1988) [113] provided stepsize rules and stopping criteria for stochastic quasigradient methods, and Ruszczynski (1987) [118] contributed by proposing a linearization method. Marti (1980) [93] improved the convergence by introducing semistochastic approximation.

King and Wets (1989)[83] have applied the theory of epi-consistency to stochastic programming in order to obtain consistency results for sequences of optimal solutions. Dupačowá and Wets (1988) [37] studied the asymptotic behavior of statistical estimators in stochastic programs.

Higle and Sen (1989) [61] developed a "Stochastic Decomposition" Benders decomposition method, which, like the stochastic quasigradient algorithm, requires only one observation or a very small number of observations per iteration, and which asymptotically creates an outer linearization of the second-stage costs. Optimality conditions and stopping rules for their method have been presented in Higle and Sen (1989) [62]. Higle, Lowe, and Odio (1990) [64] subsequently extended the framework to a multi-cut formulation.

Gaivoronski and Nazareth (1989) [52] combined generalized programming with sampling techniques. Niederreiter (1986) [102] proposed the use of pseudo random numbers for multidimensional numerical integration. Deák (1988) [33] gave a survey of well-known techniques for multidimensional integration for stochastic programming.

Monte Carlo methods are known to be efficient for multidimensional numerical integration. Lavenberg and Welch (1981) [87] and Rubinstein and Markus (1985) [116] discussed the efficiency of control variables in Monte Carlo simulation. Pereira et al. (1989) [104] used control variables as a variance-reduction technique in Monte Carlo sampling in a modified Benders decomposition framework. Dantzig and Glynn (1990) [24] and Infanger (1992) [74] used importance sampling, based on an additive approximation function, as a variance-reduction technique for Monte Carlo sampling for stochastic linear programs. Krishna (1993) [84] extended the scheme to using piecewise linear approximation functions.

Birge (1985) [10] extended the L-shaped method to multi-stage stochastic programs, employing a nested Benders decomposition scheme. Gassmann (1990) [53], based on results by Wittrock (1983) [130] for deterministic multi-stage programs, explored different tree-traversing strategies in a Benders decomposition framework for stochastic multi-stage programs. This work has been extended by Morton (1993) [94]. Louveaux (1986) [88] discussed multi-stage problems with block-separable recourse. Beale, Dantzig, and Watson (1986) [5] proposed a first-order approach to a class of multi-stage stochastic programs. Dempster (1986) [35] studied multi-stage problems. Pereira and Pinto (1991) [106] proposed stochastic dual dynamic programming, a dual Benders decomposition approach, exploiting the piecewise linear property of the recourse function, and introduced path sampling for obtaining estimates of upper bounds.

Ideas of using parallel processors can be found in Wets (1985) [127], Dantzig (1988) [23], Hillier and Eckstein (1990) [66], Zenios (1990) [131], and Ariyawansa and Hudson (1990) [2].

There has been a large variety of applications for stochastic programming; for example, Ferguson and Dantzig (1956) [47] allocated aircraft to routes, Kusy and Ziemba (1986) [85] formulated a bank asset and liability management model, Zenios (1992) [132] managed large mortgage-backed securities, Mulvey (1987) [95] formulated nonlinear networks for modeling in finance, and Pereira and Pinto (1989) [103] optimized large hydroelectric systems and (1991) [107] carried out energy planning. A description of practical models can be found in Dempster (1980) [34] and Ermoliev and Wets (1988) [46]. Overviews of stochastic programming are given in Dempster (1980) [34], Ermoliev and Wets (1988) [46], and Wets (1989) [129]. For a recent survey, see Birge and Wets (1991) [17].

2

Benders
Decomposition and
Importance Sampling

2.1   TWO-STAGE STOCHASTIC LINEAR PROGRAMS

An important class of stochastic models is that of two-stage stochastic linear programs with recourse. These models can be seen as the stochastic extensions of deterministic dynamic systems with two stages: x and y denote the first- and the second-stage decision variables; A and b represent the coefficients and right-hand sides of the first-stage constraints; and D and d concern the second-period constraints together with the transition matrix B, which couples the two periods. In the literature, D is often referred to as the technology or recourse matrix. c and f are the objective function coefficients.

In the deterministic case, c, f, A, b, B, D, and d are known with certainty to the planner. In the stochastic case, the values of the second-stage parameters are uncertain. The second-stage parameters are known at time $t = 1$ only by their probability distribution of possible outcomes; actual outcomes will be known later, at time $t = 2$. The second-stage parameters can be viewed as random variables that assume certain outcomes with certain probabilities. We denote a particular outcome of these random variables by ω and the corresponding probability by p^ω or $p(\omega)$, $\omega \in \Omega$, the set of possible outcomes. We consider the case where uncertainty occurs only in the transition matrix B and in the right-hand side vector d. The second-stage costs f and the elements of the recourse matrix D are assumed to be known with certainty.

In (2.1) a two-stage staircase problem is transformed into a two-stage stochastic linear program:

$$
\begin{array}{llll}
\min\ Z\ = & cx\ +\ E^\omega(fy^\omega) & & \\
\text{s/t} & Ax & = b & \\
& -B^\omega x\ +\ Dy^\omega & = d^\omega & \\
& x, y^\omega & \geq\ 0, & \omega \in \Omega.
\end{array}
\tag{2.1}
$$

The problem is to find a first-stage decision x that is feasible for all scenarios $\omega \in \Omega$ and has the minimum expected costs. Note the adaptive nature of the problem: while the decision x is made only with the knowledge of the distribution $(\omega, p(\omega))$ of the random parameters, the second-stage decision y^ω is made after an outcome ω is observed. The second-stage decision compensates for and adapts to different scenarios ω.

Let us consider the case of Ω being discrete and finite, and define Ω as an index set $\Omega = \{1, \ldots, K\}$, meaning that the parameter ω may take on K different values. Then we can formulate a deterministic problem that is equivalent to the stochastic linear problem. This deterministic equivalent problem is tractable only if K is small. It takes the form

$$
\begin{array}{llllllll}
\min\ Z\ = & cx & +\ p^1 fy^1 & +\ p^2 fy^2 & +\ \cdots & +\ p^K fy^K & & \\
\text{s/t} & Ax & & & & & = b & \\
& -B^1 x & +\ Dy^1 & & & & = d^1 & \\
& -B^2 x & & +\ Dy^2 & & & = d^2 & \\
& \vdots & & & \ddots & & \vdots & \\
& -B^K x & & & & +\ Dy^K & = d^K & \\
& x, & y^1, & y^2, & \ldots, & y^K & \geq\ 0. &
\end{array}
\tag{2.2}
$$

In the deterministic equivalent problem (2.2) the second-stage constraints are explicitly formulated for each scenario $\omega \in \Omega$, one below the other. The objective function carries out the expected-value computation by direct summation. Clearly, this formulation may lead to linear programs of enormous size.

2.2 BENDERS DECOMPOSITION

The method we wish to apply to solve two-stage stochastic linear programs utilizes Benders (1962) [6] decomposition. Van Slyke and Wets (1969) [122] suggested expressing the expected value of the second-stage costs by a scalar θ and replacing the second-stage constraints sequentially by "cuts," which are necessary conditions expressed only in terms of the first-stage variables x and θ. Our analysis follows this approach.

In the following we will derive the main steps of a Benders decomposition algorithm for two-stage stochastic linear programs. We will consider the "universe" case, which yields the exact solution of the equivalent deterministic problem ("certainty equivalent"). See, e.g., Geoffrion (1974) [55] for an excellent derivation of the Benders decomposition algorithm.

Given the equivalent deterministic problem in (2.2) and assuming K scenarios describe the universe case, we rewrite the problem applying projection onto the x variables and obtain (2.3). We assume for simplicity that (2.2) is feasible and has a finite optimum solution.

$$
\begin{aligned}
\min Z = \quad & \\
cx \quad + \quad & \min [p^1 fy^1 \; + \; p^2 fy^2 \; + \; \cdots \; + \; p^K fy^K] \\
Ax = b \quad & Dy^1 \qquad\qquad\qquad\qquad\qquad = d^1 + B^1 x \\
x \geq 0 \quad & \qquad\qquad Dy^2 \qquad\qquad\qquad = d^2 + B^2 x \\
& \qquad\qquad\qquad\qquad \ddots \qquad\qquad\quad \vdots \\
& \qquad\qquad\qquad\qquad\qquad Dy^K = d^K + B^K x \\
& y^1, \qquad y^2, \qquad \cdots, \qquad y^K \geq 0.
\end{aligned}
$$

$$(2.3)$$

$\min \frac{K}{i} p^i f y^i$

The infimal value function in (2.3) corresponds to the following primal linear problem (2.4):

$$
\begin{array}{llllll}
\min \ z_P \ = & p^1 f y^1 \ + & p^2 f y^2 \ + & \cdots \ + & p^K f y^K \ = & E^\omega(f y^\omega) \\
p^1 \pi^1 : & D y^1 & & & = & d^1 + B^1 x \\
p^2 \pi^2 : & & D y^2 & & = & d^2 + B^2 x \\
\vdots & & & \ddots & & \vdots \\
p^K \pi^K : & & & & D y^K \ = & d^K + B^K x \\
& y^1, & y^2, & \cdots, & y^K \ \geq & 0,
\end{array}
\tag{2.4}
$$

and to the dual linear problem (2.5):

$$
\begin{array}{l}
\max \ z_D \qquad\quad = \\
p^1 \pi^1 \ (d^1 + B^1 x) \ + \ p^2 \pi^2 \ (d^2 + B^2 x) \ + \ \cdots \ + \ p^K \pi^K \ (d^K + B^K x) \\
\quad \pi^1 \ D \hspace{8.5cm} \leq \ f \\
\qquad\qquad\qquad\quad \pi^2 \ D \hspace{5.5cm} \leq \ f \\
\hspace{7.5cm} \ddots \hspace{2.5cm} \vdots \\
\hspace{6cm} \pi^K \ D \hspace{2cm} \leq \ f.
\end{array}
\tag{2.5}
$$

The primal problem is parameterized in the right-hand side by x. The assumption that (2.2) is finite implies that (2.4) is finite for at least one value of x for which $x \geq 0$ and $Ax = b$. Applying the Duality Theorem of Linear Programming, we state that (2.5) has to be feasible. The feasibility conditions

$$
\pi^\omega D - f \leq 0
\tag{2.6}
$$

indicate that the feasible region $\{\pi^\omega | \pi^\omega D - f \leq 0\}$ is independent of x and ω and is simply repeated for each scenario $\omega \in \Omega$.

The assumption that (2.2) is feasible requires feasibility of the primal problem (2.4) for all values of x satisfying $x \geq 0$ and $Ax = b$. We define $\pi :=$ $(\pi^1, \pi^2, \ldots, \pi^K)$ to be the vector of dual variables of problem (2.5). By the Duality Theorem again, z_D in (2.5) has to be finite. Let $\pi^j, j = 1, \ldots, p$, be the extreme points and $\pi^j, j = p+1, \ldots, p+q$, be representatives of the extreme rays of the feasible region of (2.5), where $\pi^j := (\pi^{1j}, \pi^{2j}, \ldots, \pi^{Kj})$. Problem (2.5) is finite if and only if

$$\pi^{\omega j}(d^\omega + B^\omega x) \leq 0, \quad j = p+1, \ldots, p+q, \ \omega \in \Omega. \tag{2.7}$$

Constraints (2.7) may be appended to problem (2.3) to ensure that the dual problem (2.5) is bounded.

Next we outer linearize the infimal value function in (2.3), whose value is exactly

$$\max_{j=1,\ldots,p} \sum_{\omega \in \Omega} p^\omega \pi^{\omega j}(d^\omega + B^\omega x). \tag{2.8}$$

By expressing the infimal value function by the outer linearized dual problem and using θ as the smallest upper bound, we can represent the problem in the following form:

$$\begin{aligned} \min \ Z = cx \ &+ \ \theta \\ Ax \ &= \ b \\ x \ &\geq \ 0 \end{aligned} \tag{2.9}$$

$$\begin{aligned} \theta &\geq \textstyle\sum_{\omega \in \Omega} p^\omega \pi^{\omega j}(d^\omega + B^\omega x), && j = 1, \ldots, p \\ \pi^{\omega j}(d^\omega + B^\omega x) &\leq 0, && j = p+1, \ldots, p+q, \ \omega \in \Omega. \end{aligned} \tag{2.10}$$

Relaxation is applied to represent constraints (2.10) because we do not want to know all $\pi^j, j = 1, \ldots, p+q$ in advance. Given a solution $(\hat{x}, \hat{\theta})$ from the master problem, one solves the separable problem (2.4) or problem (2.5) by solving the individual problems (2.11):

$$
\begin{aligned}
z^{\omega*}(\hat{x}) = \min \ z_P^{\omega} \ &= \ fy^{\omega} \\
D y^{\omega} \ &= \ d^{\omega} + B^{\omega} x \\
y^{\omega} \ &\geq \ 0, \qquad \omega \in \Omega,
\end{aligned}
\tag{2.11}
$$

or the dual problems (2.12) of these:

$$
\begin{aligned}
z^{\omega*}(\hat{x}) = \max z_D^{\omega} \ &= \ \pi^{\omega}(d^{\omega} + B^{\omega}\hat{x}) \\
\pi^{\omega} D \ &\leq \ f, \qquad \omega \in \Omega.
\end{aligned}
\tag{2.12}
$$

We call $\pi^{\omega*}(\hat{x})$ the optimum dual solution vector. If primal infeasibility (dual unboundedness) is detected, with $\pi^{\omega^{\circ}}(\hat{x})$ denoting the corresponding extreme ray, a feasibility cut

$$
\pi^{\omega^{\circ}}(\hat{x}) \cdot (d^{\omega} + B^{\omega} x) \leq 0
\tag{2.13}
$$

is added to the master problem. If all primal problems are feasible (all dual problems bounded), an optimality cut

$$
\theta \geq \sum_{\omega \in \Omega} p^{\omega} \pi^{\omega*}(\hat{x}) \cdot (d^{\omega} + B^{\omega} x)
\tag{2.14}
$$

is added to the master problem. We call

$$
L(\hat{x}, x) := \sum_{\omega \in \Omega} p^{\omega} \pi^{\omega*}(\hat{x}) \cdot (d^{\omega} + B^{\omega} x)
\tag{2.15}
$$

an outer linearization of the second-stage costs, which are defined by

$$z(\hat{x}) := \sum_{\omega \in \Omega} p^{\omega} z^{\omega*}(\hat{x}). \tag{2.16}$$

The relation

$$L(\hat{x}, x) \le z(x) \tag{2.17}$$

formulates the main property of the outer linearization. Any cut regardless of the $\hat{x}$ from which it was originally derived is a valid cut as long as it does not violate the main property of outer linearization.

Benders decomposition provides upper and lower bounds for the solution in each iteration. In the lth iteration,

$$\text{LB}^{l} := c\hat{x}^{l} + \hat{\theta}^{l}, \tag{2.18}$$

with $\hat{x}^{l}$, $\hat{\theta}^{l}$ being the optimum solution of the master problem in iteration l, is defined to be a lower bound, and

$$\text{UB}^{l} := \min\{\text{UB}^{l-1}, c\hat{x}^{l} + z(\hat{x}^{l})\}, \quad \text{UB}^{0} = \infty, \tag{2.19}$$

with $z(\hat{x}^{l})$ being the second-stage costs associated with the solution $\hat{x}^{l}$ of the master problem, is defined to be an upper bound to the solution of the problem. If

$$(\text{UB}^{l} - \text{LB}^{l})/\text{LB}^{l} \le \text{TOL}, \tag{2.20}$$

where TOL is a given tolerance, the problem is said to be solved with a sufficient accuracy.

Summarizing, Benders decomposition splits the original problem into a master problem and a subproblem, which in turn decomposes into a series of independent subproblems, one for each $\omega \in \Omega$. The latter are used to generate cuts. The master problem, the subproblems, and the cuts are summarized in (2.21), (2.22), and (2.23).

The master problem:

$$
\begin{aligned}
\min \ z_M \ &= \ && cx \ + \ && \theta \\
\text{s/t} \quad && Ax \ && &= \ b \\
&& -G^l x \ + \ && \alpha^l \theta \ &\geq \ g^l, \ l = 1, \ldots, L \\
&& x, \ && \theta \ &\geq \ 0
\end{aligned}
\tag{2.21}
$$

The subproblems:

$$
\begin{aligned}
\min \ z^\omega \ &= \ fy^\omega \\
\text{s/t} \ \pi^\omega : \quad Dy^\omega \ &= \ d^\omega + B^\omega x \\
y^\omega \ &\geq \ 0, \ \omega \in \Omega, \ \text{e.g.,} \ \Omega = \{1, 2, \ldots, K\},
\end{aligned}
\tag{2.22}
$$

where $\pi^{\omega*}$ is the optimal dual solution of subproblem ω.

The cuts:

$$
\begin{aligned}
g^l \ &= \ \sum_\omega p^\omega \pi^{\omega*}(\hat{x}^l) d^\omega \ = \ E \, \pi^{\omega*}(\hat{x}^l) d^\omega, \\
G^l \ &= \ \sum_\omega p^\omega \pi^{\omega*}(\hat{x}^l) B^\omega \ = \ E \, \pi^{\omega*}(\hat{x}^l) B^\omega,
\end{aligned}
\tag{2.23}
$$

$\alpha^l = 0$ for a feasibility cut, $\alpha^l = 1$ for an optimality cut. $\qquad$ (2.24)

By solving the master problem (2.21), where cuts are initially absent and then sequentially added, we obtain a trial solution $\hat{x}^l$. Given $\hat{x}^l$ we can solve K subproblems $\omega \in \Omega$ (2.22) to compute a cut (2.23). The cut is a supporting hyperplane for the expected value of the second-stage costs represented as a function of x. Cuts are sequentially added to the master problem, and new values of $\hat{x}^l$ are obtained until the optimality criterion is met. We distinguish between two types of cuts: feasibility cuts and optimality cuts. The first type refers to infeasible subproblems for a given $\hat{x}^l$ and the latter to feasible and optimum subproblems, given $\hat{x}^l$.

If the expected values z, G, and g are computed exactly, that is, by evaluating all scenarios $\omega \in \Omega$, we refer to this case as the universe case. As we will see later, the number of scenarios easily gets out of hand, and it is not always possible to solve the universe case. Methods are therefore sought that guarantee a satisfactory solution without having to solve the universe case. We employ Monte Carlo sampling techniques to obtain accurate estimates of the expected values z, G, and g.

2.3 MONTE CARLO SAMPLING

2.3.1 Multidimensional Integration

The computation of the expected future costs z and of the coefficients G and right-hand side g of the cuts requires the computation of multiple integrals or multiple sums. The expected value of the second-stage costs, e.g., $z = E\ z^\omega = E(C)$, is an expectation of functions $C(v^\omega)$, $\omega \in \Omega$, where $C(v^\omega)$ is obtained by solving a linear program. V (in general) is an h-dimensional random vector parameter, $V = (V_1, \ldots, V_h)$, with outcomes $v^\omega = (v_1, \ldots, v_h)^\omega$. An outcome of v^ω determines a particular value of B^ω and b^ω. For example, in expansion-planning problems of electric power systems, V_i represents the percentage of generators of type i down for repair or transmission lines of type i not operating, and v_i^ω is the observed random percentage outcome. In portfolio management problems, V_i might represent the random value of the independent factor i used to describe the random value of returns of the universe of assets in a portfolio, and v_i^ω the observed outcome. The vector v^ω is also denoted by v, and $p(v^\omega)$, alias $p(v)$, denotes the corresponding probability. Ω is the set of all possible random events and is constructed by crossing the sets of outcomes: $\Omega = \Omega_1 \times \Omega_2 \times \cdots \times \Omega_h$. We assume independence of the stochastic parameters $V_1, \ldots, V_h$. With P being the probability measure, $E\,C(V)$ takes the form of a multiple integral $E\,C(V) = \int C(v^\omega)P(d\omega) = \int \cdots \int C(v)p(v)\,dv_1 \cdots dv_h$, or, in the case of discrete distributions, the form of a multiple sum $E\,C(V) = \sum_{v_1} \cdots \sum_{v_h} C(v)p(v)$. Based on the independence assumption, $p(v) = p_1(v_1) \cdots p_h(v_h)$.

In the following discussion we concentrate on discrete distributions. This is not a restriction, as the approach can be easily adapted for continuous distributions. In practical applications we may assume that all distributions can be approximated with sufficient accuracy by discrete ones. In practical problems the number of terms in the multiple-sum computation becomes astronomically large and therefore the evaluation of multiple sums by direct summation is not practical. This is especially true because the evaluation of each term in the multiple sum requires the solution of a linear program. For example, in order to compute the expected values of the subproblem costs, and of the coefficients and right-hand sides of the cuts, a linear program has to be solved for each outcome $\omega \in \Omega$.

The expected value of the subproblem costs is denoted by

$$z = E\,C(v^\omega) = E\,fy^{*\omega}, \quad \omega \in \Omega, \tag{2.25}$$

with $y^{*\omega}$ being the optimum solution of subproblem ω. The number of elements of Ω is determined by the dimensionality of the stochastic vector $V = (V_1, \ldots, V_h)$. Typically the dimension h of V is quite large.

For example, in expansion-planning problems of electric power systems, one component of V denotes the availability of one type of generator or one demand of power in a node of a multiarea supply network, or the availability of one type of transmission line connecting two nodes. Consider several nodes and arcs and one demand and some options of generators at each node. The number of scenarios K in the universe case quickly gets out of hand, even if the distribution of each component of V is determined by just a small number K^i of discrete points. Suppose, e.g., that $h = 20$ and $K^i = 2$, $i = 1, \ldots, 20$. This means that

20 components of the system are either in operation or down for repair. The resulting number of scenarios is as large as $2^{20} \approx 10^6$. Alternatively, suppose that $h = 20$ and $K^i = 5$, $i = 1, \ldots, 20$. Then the total number of terms in the expected value calculations is $K = 5^{20} \approx 10^{14}$, which is not practically solvable by direct summation.

In portfolio management problems, one component of V may denote the value of an orthogonal factor. Usually up to 80 factors are used to describe the random outcomes of up to several thousand assets. Suppose, for example, we used 80 factors and discrete stochastic parameters with only three outcomes each. Then the number of possible scenarios would be as much as $3^{80} \approx 10^{38}$, an astronomically large number.

Monte Carlo methods appear promising for the computation of multiple integrals or multiple sums for large h (Davis and Rabinowitz (1984) [32]). See Hammersly and Handscomb (1964) [60] for a description of Monte Carlo sampling techniques.

2.3.2 Crude Monte Carlo

Suppose $v^\omega, \omega = 1, \ldots, N$, are scenarios sampled independently from their joint probability mass function. Then $C^\omega = C(v^\omega)$ are independent random variates with expectation z.

$$\bar{z} = (1/N) \sum_{\omega=1}^{N} C^\omega \tag{2.26}$$

is an unbiased estimator of z, and its variance is

$$\sigma_{\bar{z}}^2 = \sigma^2/N, \tag{2.27}$$

where $\sigma^2 = \text{var}(C(V))$. Thus the standard error decreases with sample size N by $N^{-0.5}$. The convergence rate of $\bar{z}$ to z is independent of the dimension h of the random vector V.

2.3.3   Importance Sampling

We rewrite

$$z = \sum_{\omega \in \Omega} C(v^\omega)p(v^\omega) = \sum_{\omega \in \Omega} \frac{C(v^\omega)p(v^\omega)q(v^\omega)}{q(v^\omega)} \qquad (2.28)$$

by introducing a probability mass function $q(v^\omega)$. We can view q as a probability mass function of a random vector W that assumes the same outcomes as V but with different probabilities. Therefore, by change of variables,

$$z = E\frac{C(W)p(W)}{q(W)}. \qquad (2.29)$$

We obtain a new estimator of z,

$$\bar{z} = \frac{1}{N} \sum_{\omega=1}^{N} \frac{C(w^\omega)p(w^\omega)}{q(w^\omega)}, \qquad (2.30)$$

which has a variance of

$$\mathrm{var}(\bar{z}) = \frac{1}{N} \sum_{\omega \in \Omega} \left(\frac{C(w^\omega)p(w^\omega)}{q(w^\omega)} - z \right)^2 q(w^\omega). \qquad (2.31)$$

Choosing $q^*(w^\omega) = C(w^\omega)p(w^\omega)/\sum_{\omega \in \Omega} C(w^\omega)p(w^\omega)$ would lead to $\mathrm{var}(\bar{z}) = 0$, which means we could get a perfect estimate of the multiple sum by just one single observation. However, this is practically useless, since to sample Cp/q we have to know q and to determine q we need to know $z = \sum_{\omega \in \Omega} C(w^\omega)p(w^\omega)$, which we eventually want to compute. This result nevertheless helps to derive heuristics for how to choose q: it should be approximately proportional to the product $C(w^\omega)p(w^\omega)$ and have a form that can be integrated analytically. For instance, using the additive (separable in the components of the stochastic vector) approximation of (2.32),

$$C(V) \approx \sum_{i=1}^{h} C_i(V_i), \tag{2.32}$$

could be a possible way to compute a proper q:

$$q(w^\omega) \approx \frac{C(w^\omega)p(w^\omega)}{\sum_{i=1}^{h} \sum_{\omega \in \Omega_i} C_i(w^\omega)p_i(w^\omega)}. \tag{2.33}$$

In this case one has to evaluate only h one-dimensional sums instead of one h-dimensional sum. Depending on how well the additive model approximates the original cost surface, the above-mentioned estimator will lead to smaller variances than crude Monte Carlo sampling. Of course, if the original cost surface has the property of additivity (separability), no sampling is required, as the multiple sum is computed exactly by h one-dimensional sums.

The advantage of this approach lies in the fact that even if the additive model is a poor approximation to the cost surface, the method still works. The price that has to be paid is a high sample size. The variance reduction compared to crude Monte Carlo will be small. For the theory of importance sampling we refer to Glynn and Iglehart (1989) [58]. See also Dantzig and Glynn (1990) [24].

R. Entriken and M. Nakayama in Dantzig et al. (1989) [25] developed an importance sampling scheme using an additive model to approximate the cost function $E\,C(V)$. In particular, $C(v)$ is approximated by a marginal cost model, considering marginal costs in each dimension i of V and a base case, the point from which the approximation is developed. We will use this approach here. As we employ importance sampling within the Benders decomposition algorithm, the costs $C(v, \hat{x})$, the approximation of the costs $\Gamma(v, \hat{x})$, and thus the importance distribution of $q(v, \hat{x})$ depend also on $\hat{x}$, the current solution of the master problem. Introducing the costs of the base case $C(\tau, \hat{x})$ makes the model more

[handwritten margin note: Key connexion between Benders & importance sampling]

sensitive to the impact of the stochastic variables V:

$$C(V, \hat{x}) \approx \Gamma(V, \hat{x}) = C(\tau, \hat{x}) + \sum_{i=1}^{h} M_i(V_i, \hat{x}), \qquad (2.34)$$

$$M_i(V_i, \hat{x}) = C(\tau_1, \ldots, \tau_{i-1}, V_i, \tau_{i+1}, \ldots, \tau_h, \hat{x}) - C(\tau, \hat{x}). \qquad (2.35)$$

Marginal cost in i^{th} variable

The vector $\tau = (\tau_1, \ldots, \tau_h)$ can be any arbitrarily chosen point out of the set of values v_i, $i = 1, \ldots, h$. For example, we choose τ_i as the outcome of V_i that leads to the lowest costs. Figure 2.1 represents schematically the true cost function $C(V)$ and the additive approximation function $\Gamma(V)$ for the case of only two dimensions.

Note that the second-stage costs are computed by a linear program, where the uncertain parameters appear on the right-hand side. Therefore the second-stage costs are convex in the stochastic parameters V. In the most general case of a convex function $C(V_1, \ldots, V_h, \hat{x})$, the choice of the base case $\tau = (\tau_1, \ldots, \tau_h)$ to be the values of $(V_1, \ldots, V_h)$ that minimize the function $C(V_1, \ldots, V_h, \hat{x})$ for given $\hat{x}$ requires the solution of a convex minimization problem (or a discrete optimization problem in the case of discrete distributions of V). This is impractical, and another choice of a base case is taken. In many cases the base case of lowest costs can be easily found. For example, in the context of expansion planning of power systems, choosing the base case of lowest costs means selecting respectively lowest demands and highest availabilities of generators and transmission lines.

Defining

expected i^{th} marginal cost

$$\bar{M}_i(\hat{x}) = E\, M_i(V_i, \hat{x}) = \sum_{\omega \in \Omega_i} M_i(v_i^{\omega}, \hat{x}) p(v_i^{\omega}) \qquad (2.36)$$

and

$$F(v^{\omega}, \hat{x}) = \frac{C(v^{\omega}, \hat{x}) - C(\tau, \hat{x})}{\sum_{i=1}^{h} M_i(v_i^{\omega}, \hat{x})}, \qquad (2.37)$$

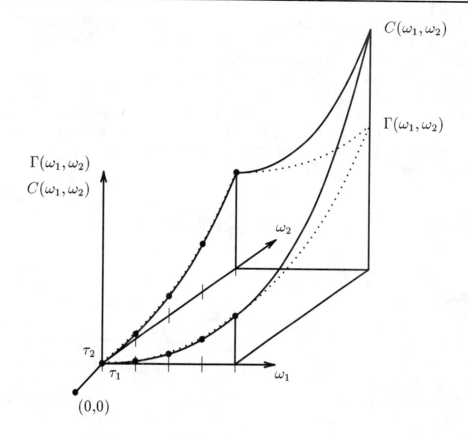

Figure 2.1: Additive approximation versus true cost function

where we assume that $\sum_{i=1}^{h} M_i(v_i^\omega, \hat{x}) > 0$, so that at least one $M_i(v_i^\omega, \hat{x}) > 0$, we can express the expected value of the costs in the following form:

$$z(\hat{x}) = C(\tau, \hat{x}) + \sum_{i=1}^{h} \bar{M}_i(\hat{x}) \sum_{\omega \in \Omega} F(v^\omega, \hat{x}) \frac{M_i(v_i^\omega, \hat{x})}{\bar{M}_i(\hat{x})} \prod_{j=1}^{h} p_j(v_j^\omega). \tag{2.38}$$

Note that this formulation consists of a constant term and a sum of h expectations. Given a fixed sample size N, we partition N into h subsamples, with sample sizes N_i, $i = 1, \ldots, h$, such that $\Sigma N_i = N$ and $N_i \geq 1$, $i = 1, \ldots, N$, where N_i is approximately proportional to $\bar{M}_i$. The h expectations are separately approximated by sampling marginal densities. The ith expectation corresponds to the ith component of V. In generating sample points for the ith expectation, we use the importance density $(p_i M_i / \bar{M}_i)$ for sampling the ith component of V and the original marginal densities for any other components. Denoting the estimate of the ith sum by

$$\mu_i(\hat{x}) = \frac{1}{N_i} \sum_{j=1}^{N_i} F(v^j, \hat{x}), \tag{2.39}$$

we obtain

$$\bar{z}(\hat{x}) = C(\tau, \hat{x}) + \sum_{i=1}^{h} \bar{M}_i(\hat{x}) \mu_i(\hat{x}), \tag{2.40}$$

the estimated expected value of the second-stage costs.

Let $\bar{\sigma}_i^2(\hat{x})$ be the sample variance of the ith expectation, where $\sigma_i^2(\hat{x}) = 0$ if $N_i = 1$. The estimated variance of the mean, $\sigma_{\bar{z}}^2(\hat{x})$, is then given by

$$\sigma_{\bar{z}}^2(\hat{x}) = \sum_{i=1}^{h} \frac{\bar{M}_i^2(\hat{x}) \bar{\sigma}_i^2(\hat{x})}{N_i}. \tag{2.41}$$

Using importance sampling, one can achieve significant variance reduction. The experiment of M. Nakayama in Dantzig et al. (1989) [25] claims a variance reduction of 1:20,000 using importance sampling versus crude Monte Carlo sampling. For a given optimal $\hat{x}$, the second-stage costs of a multiarea expansion-planning model with 192 universe scenarios were sampled with a sample size of 10 using both methods, and the results were compared.

The derivation above concerned the estimation of the expected second-stage costs $z(\hat{x})$. To derive a cut, we use an analogous framework. Note that a cut is defined as an outer linearization of the second-stage costs represented as a function of x, the first-stage variables. At $\hat{x}$, the value of the cut is exactly the expected second-stage costs $z(\hat{x})$. Note also that any choice of q is a valid choice and leads to an unbiased estimate. As we do not want to derive different importance distributions for the coefficients and the right-hand side of a cut, we use the q already at hand from the cost estimation. We therefore employ directly the cost approximation scheme and the importance distribution to compute the gradient and the right-hand side of a cut. With $B(v^\omega) := B^\omega$ and $d(v^\omega) := d^\omega$ as the outcomes of B and d in scenarios $\omega \in \Omega$ and $\pi^*(v^\omega, \hat{x}) := \pi^{\omega*}(\hat{x})$ the optimum dual solution in scenario ω, we define

$$F^G(v^\omega, \hat{x}) = \frac{\pi^*(v^\omega, \hat{x})B(v^\omega) - \pi^*(\tau, \hat{x})B(\tau)}{\sum_{i=1}^{h} M_i(v_i^\omega, \hat{x})}, \qquad (2.42)$$

$$F^g(v^\omega, \hat{x}) = \frac{\pi^*(v^\omega, \hat{x})d(v^\omega) - \pi^*(\tau, \hat{x})d(\tau)}{\sum_{i=1}^{h} M_i(v_i^\omega, \hat{x})}, \qquad (2.43)$$

and compute

$$G(\hat{x}) = \pi^*(\tau, \hat{x})B(\tau) + \sum_{i=1}^{h} \bar{M}_i(\hat{x}) \sum_{\omega \in \Omega} F^G(v^\omega, \hat{x}) \frac{M_i(v_i^\omega, \hat{x})}{\bar{M}_i(\hat{x})} \prod_{j=1}^{h} p_j(v_j^\omega), \qquad (2.44)$$

$$g(\hat{x}) = \pi^*(\tau, \hat{x})d(\tau) + \sum_{i=1}^{h} \bar{M}_i(\hat{x}) \sum_{\omega \in \Omega} F^g(v^\omega, \hat{x}) \frac{M_i(v_i^\omega, \hat{x})}{\bar{M}_i(\hat{x})}) \prod_{j=1}^{h} p_j(v_j^\omega), \qquad (2.45)$$

the coefficients and the right-hand side of a cut. We estimate the expected values again by sampling, using the sample points at hand from the computation of $\bar{z}$.

Using Monte Carlo sampling, we obtain $\bar{z}(\hat{x})$, $\bar{G}(\hat{x})$, and $\bar{g}(\hat{x})$, which are approximations of the expected values $z(\hat{x})$, $G(\hat{x})$, and $g(\hat{x})$. We also obtain the

Do use sample subspace $\hat{\Omega}$ in 2ⁿᵈ sum in (2.44), (2.45)

estimated variance of the mean of the second-stage costs $\sigma_{\bar{z}}^2(\hat{x})$. The impact of using approximations instead of the exact parameters on the Benders decomposition algorithm is analyzed in the following section.

2.4 PROBABILISTIC CUTS

2.4.1 The Influence of the Estimation Error

Employing Monte Carlo sampling techniques means not solving all problems $\omega \in \Omega$, but solving problems $\omega \in S$, S being a subset of Ω. Instead of the exact expected values $z(\hat{x})$, $G(\hat{x})$, and $g(\hat{x})$, we compute the estimates $\bar{z}(\hat{x})$, $\bar{G}(\hat{x})$, and $\bar{g}(\hat{x})$ by importance sampling. We also estimate the error of the estimation of $z(\hat{x})$ by the variance $\mathrm{var}(\bar{z}(\hat{x})) = \sigma_{\bar{z}}^2(\hat{x})$. Thus, in the case of the second-stage costs, for example, the estimation results in an estimated mean with some error distribution. There is good reason to assume that the error is normally distributed (Davis and Rabinowitz (1984) [32]). We define $\tilde{z}(\hat{x})$ to be normally distributed with mean $\bar{z}(\hat{x})$ and variance $\sigma_{\bar{z}}^2(\hat{x})$:

$$\tilde{z}(\hat{x}) := N(\bar{z}(\hat{x}), \sigma_{\bar{z}}^2(\hat{x})). \tag{2.46}$$

A cut obtained by sampling differs in general from a cut computed by solving the universe scenarios. The outer linearizations

$$L(\hat{x}, x) = G(\hat{x})x + g(\hat{x}), \tag{2.47}$$

with respect to the universe case, and

$$\bar{L}(\hat{x}, x) = \bar{G}(\hat{x})x + \bar{g}(\hat{x}), \tag{2.48}$$

with respect to the estimation, differ in the gradient and the right-hand side. We will denote this difference as $\epsilon_{\hat{x}}(x)$; thus,

$$\epsilon_{\hat{x}}(x) = L(\hat{x}, x) - \bar{L}(\hat{x}, x). \tag{2.49}$$

At $x = \hat{x}$, the value at which the cut was derived, $L(\hat{x}, \hat{x}) = z(\hat{x})$ and $\bar{L}(\hat{x}, \hat{x}) = \bar{z}(\hat{x})$.

Thus, if a true cut obtained by solving the universe case is binding at the solution $x = \hat{x}$, the variable θ takes on the value

$$\theta = L(\hat{x}, \hat{x}) = z(\hat{x}). \tag{2.50}$$

In the case of using Monte Carlo sampling, we assign θ to be the value of the estimated expected costs at $\hat{x}$, $\bar{L}(\hat{x}, \hat{x}) = \bar{z}(\hat{x})$, and correct for the estimation error through the right-hand side. Thus

$$\theta = \bar{L}(\hat{x}, \hat{x}) - \bar{z}(\hat{x}) + z(\hat{x}), \tag{2.51}$$

which we write as

$$\theta = \bar{L}(\hat{x}, \hat{x}) + \epsilon_{\hat{x}}(\hat{x}). \tag{2.52}$$

Equation (2.51) represents a valid statement for a solution $x = \hat{x}$. The correction term $\epsilon_{\hat{x}}(\hat{x}) = z(\hat{x}) - \bar{z}(\hat{x})$ corrects for the estimation error. Of course, we do not know the difference $z(\hat{x}) - \bar{z}(\hat{x})$ explicitly for each cut when we compute it. However, we can obtain an estimate of the distribution of $\epsilon_{\hat{x}}(\hat{x})$ by the estimation process. Using Monte Carlo importance sampling, we obtain an unbiased estimate of $z(\hat{x})$, $\bar{z}(\hat{x})$, with variance $\sigma_{\bar{z}}^2(\hat{x})$. Thus $\epsilon_{\hat{x}}(\hat{x})$ is normally distributed with mean 0 and variance $\sigma_{\bar{z}}^2(\hat{x})$:

$$\epsilon_{\hat{x}}(\hat{x}) := N(0, \sigma_{\bar{z}}^2(\hat{x})). \tag{2.53}$$

Suppose next that a cut $\bar{L}(\hat{x}, x) = \bar{G}(\hat{x})x + \bar{g}(\hat{x})$ is computed at $x = \hat{x}$, but is binding at a solution $\hat{\hat{x}}$ where $\hat{x} \neq \hat{\hat{x}}$. Applying the correction for the estimation error resulting from using Monte Carlo sampling instead of computing the expectations exactly, we obtain

$$\theta = \bar{L}(\hat{x}, \hat{\hat{x}}) + \epsilon_{\hat{x}}(\hat{\hat{x}}). \tag{2.54}$$

The correction term for the estimation error is clearly the true value of the cut at $x = \hat{\hat{x}}$ (obtained by solving the universe case) minus the value obtained by the sampling procedure:

$$\epsilon_{\hat{x}}(\hat{\hat{x}}) = L(\hat{x}, \hat{\hat{x}}) - \bar{L}(\hat{x}, \hat{\hat{x}}). \tag{2.55}$$

Again we do not know the difference $L(\hat{x}, \hat{\hat{x}}) - \bar{L}(\hat{x}, \hat{\hat{x}})$ when we compute the cut. Also, the distribution of the estimation error $\epsilon_{\hat{x}}(\hat{\hat{x}})$ at $x = \hat{\hat{x}}$ is not known directly by the estimation procedure but can be computed. Thus, initially, and sufficiently for practical purposes, we assume that

$$\epsilon_{\hat{x}}(\hat{\hat{x}}) \approx \epsilon_{\hat{x}}(\hat{x}) \quad \text{for} \quad \hat{\hat{x}} \approx \hat{x}. \tag{2.56}$$

This implies that the error distribution $\epsilon_{\hat{x}}(x)$ is assumed to be approximately constant with respect to x. We will show empirically that this assumption is valid for most practical problems. One can intuitively see it is applicable because we can expect that cuts will be binding at an x very close to the $\hat{x}$ at which they were originally derived, as the value of x changes little before we terminate at the final solution. Note that, in general, the set of observations S is a sufficiently large subset of Ω so that the variance $\sigma_{\bar{z}}^2$ is small.

Note that cuts computed by sampling do not necessarily meet the condition of outer linearization. They may intersect and separate parts of the feasible region of the second-stage problem. A sampled cut is therefore not necessarily a valid cut. The correction terms $\epsilon_{\hat{x}}(\hat{x})$ and $\epsilon_{\hat{x}}(x)$, however, account for the error we make when using Monte Carlo sampling instead of solving the universe case.

2.4.2 The Estimation Error of a Cut as a Function of x

Using Monte Carlo sampling, we estimate the coefficients $G_i, i = 1, \ldots, n$, and the right-hand side g and obtain $\bar{G}$ and $\bar{g}$ by averaging over $G^\omega = G_1^\omega, \ldots, G_n^\omega$ and g^ω. For example, in the case of crude Monte Carlo sampling, a sample of size N would be obtained by sampling G^ω and g^ω, $\omega = 1, \ldots, N$, from the original distribution $p(\omega)$, $\omega \in \Omega$, and an estimate of G and g would be computed by the sample means $\bar{G}$ and $\bar{g}$. In the case of importance sampling, a different and more complicated weighting scheme is used. We do not discuss this here; we continue with the case of crude Monte Carlo.

For each sample ω, $\omega = 1, \ldots, N$, we obtain independent observations of the mutually dependent coefficients and right-hand side of a cut:

$$G_1^\omega, \ldots, G_n^\omega, g^\omega. \tag{2.57}$$

Let us denote by R the $n + 1 \times N$ matrix composed of the observations of the coefficients and the right-hand side of the cuts, adjusted by the mean of the observations:

$$R = \begin{pmatrix} G_1^1 - \bar{G}_1 & \cdots & G_1^N - \bar{G}_1 \\ \vdots & & \vdots \\ G_n^1 - \bar{G}_n & \cdots & G_n^N - \bar{G}_n \\ g^1 - \bar{g} & \cdots & g^N - \bar{g} \end{pmatrix}. \tag{2.58}$$

An estimate of the variance of the mean value of $\bar{L}(\hat{x}, x) = \bar{G}(\hat{x})x + \bar{g}(\hat{x})$ can be obtained by

$$\mathrm{var}(\bar{L}(\hat{x}, x)) = \frac{1}{N} \frac{1}{N-1} (x^T, 1) R R^T (x^T, 1)^T. \tag{2.59}$$

Defining

$$u^\omega = R^\omega (x^T, 1)^T, \quad \omega = 1, \ldots, N, \tag{2.60}$$

where the row vector R^ω extracts the column in R that represents the observation ω,

$$R^\omega = (G_1^\omega - \bar{G}_1, \ldots, G_n^\omega - \bar{G}_n, g^\omega - \bar{g}), \tag{2.61}$$

we can write the estimated variance of the value of the cut at x as

$$\text{var}(\bar{L}(\hat{x}, x)) = \frac{1}{N} \frac{1}{N-1} \sum_{\omega=1}^{N} (u^{\omega})^2. \tag{2.62}$$

This estimate can be easily computed. It follows that, with knowledge of the matrix R of observations of the coefficients and the right-hand side of a cut, we are able to compute an estimate of the variance of the value of the cut, $\text{var}(\bar{L}(\hat{x}, x))$, for any value of x. Clearly,

$$\text{var}(\bar{L}(\hat{x}, \hat{x})) = \text{var}(\bar{z}(\hat{x})). \tag{2.63}$$

Using the estimation procedure, we obtain an estimate of the distribution of the correction term:

$$\epsilon_{\hat{x}}(x) = N(0, \text{var}(\bar{L}(\hat{x}, x))), \tag{2.64}$$

where we also denote

$$\sigma_{\bar{L}}^2(x) := \text{var}(\bar{L}(\hat{x}, x)). \tag{2.65}$$

The error term $\epsilon_{\hat{x}}(x)$ now correctly represents an estimate of the distribution of the difference $L(\hat{x}, x) - \bar{L}(\hat{x}, x)$ as a function of x.

It seems impractical to store the matrix R to obtain the error estimate of each cut as a function of a particular solution of the master problem at which the cut is binding, but a Taylor approximation of $\text{var}(\bar{L}(\hat{x}, x))$ seems to be computationally advantageous.

Let $\alpha(\hat{x})$ be the vector of local derivatives at $\hat{x}$:

$$\alpha(\hat{x}) := \left[\frac{\partial}{\partial x} (\text{var}(\bar{L}(\hat{x}, x))) \right]_{\hat{x}}, \tag{2.66}$$

where

$$\frac{\partial}{\partial x_i}(\text{var}(\bar{L}(\hat{x},x))) = \frac{\partial}{\partial x_i}\left(\frac{1}{N}\frac{1}{N-1}\sum_{\omega=1}^{N}(u^\omega)^2\right) = \frac{2}{N(N-1)}\sum_{\omega=1}^{N}R^\omega(x^T,1)^T R_i^\omega.$$

(2.67)

Using a first-order Taylor approximation, we can write $\text{var}(\bar{L}(\hat{x},x))$ in the following form:

$$
\begin{aligned}
\sigma_{\bar{L}}^2(x) = \text{var}(\bar{L}(\hat{x},x)) &\approx \text{var}(\bar{L}(\hat{x},\hat{x})) &+& \alpha(\hat{x})(x-\hat{x}) \\
&= \text{var}(\bar{z}(\hat{x})) &+& \alpha(\hat{x})(x-\hat{x}).
\end{aligned}
$$

(2.68)

The estimate of the variance $\text{var}(\bar{L}(\hat{x},x))$ as a function of x is represented as a sum of two terms: the estimated variance of the second-stage costs at $\hat{x}$, $\text{var}(\bar{z}(\hat{x}))$, plus the linear term that represents the change of the variance, for $x \neq \hat{x}$. The latter term is assumed to be small, as $\alpha(\hat{x})$ is assumed to be small.

We will in the following denote the constant error term $\epsilon_{\hat{x}}(\hat{x})$ as ϵ or as ϵ^l, short for $\epsilon_{\hat{x}^l}(\hat{x}^l)$, when we refer to it in a particular iteration l of the Benders decomposition algorithm. We will denote the error term as a function of x, $\epsilon_{\hat{x}}(x)$, as $\epsilon(x)$ or as $\epsilon^l(x)$, short for $\epsilon_{\hat{x}^l}(x)$, when referring to it in a particular Benders iteration l. We will continue with the assumption that the error is approximately constant with respect to x. That means we see the error concentrated mainly in the right-hand side of the cut, and we assign the constant error term ϵ rather than the variable correction term $\epsilon(x)$ to the right-hand side. We then take

$$-\bar{G}x + \theta \geq \bar{g} + \epsilon$$

(2.69)

to be approximately a valid cut. However, everything that will be derived using the constant error term ϵ can be extended to using the variable error term $\epsilon(x)$. We will show this at the relevant places in the text.

2.4.3   Upper and Lower Bounds

For random second-stage costs $\tilde{z}(\hat{x}^l)$ and random right-hand sides $\bar{g}^l + \epsilon^l$, $l = 1, \ldots, L$, the upper and lower bounds of the problem as provided by the Benders decomposition algorithm are probabilistic.

The Upper Bounds

$$\tilde{\mathrm{UB}}^l := c\hat{x}^l + \tilde{z}(\hat{x}^l), \quad l = 1, \ldots, L, \tag{2.70}$$

are random parameters, normally distributed with means $\bar{\mathrm{UB}}^l$ and variances $\sigma_{\tilde{z}}^2(\hat{x}^l)$:

$$\tilde{\mathrm{UB}}^l := N(\bar{\mathrm{UB}}^l, \sigma_{\tilde{z}}^2(\hat{x}^l)), \quad l = 1, \ldots, L. \tag{2.71}$$

We define the lowest upper bound to be the upper bound with the lowest mean:

$$\tilde{\mathrm{UB}}_{\min}^L := \tilde{\mathrm{UB}}_{\min}^k, \quad k \in \arg\min\{\bar{\mathrm{UB}}^l,\ l = 1, \ldots, L\}, \tag{2.72}$$

with corresponding variance $\sigma_{\mathrm{UB}_{\min}^L}^2$, as estimated from the sampling process.

The Lower Bounds

The lower bounds are obtained from the solution of the master problem. To determine the distribution of a lower bound, consider the master problem at iteration L:

$$
\begin{aligned}
\tilde{\mathrm{LB}}^L \;=\; \tilde{z}_M^{*L} \;=\; \min\ \tilde{z}_M^L \;=\;\quad & cx \;+\; \theta \\
\text{s/t}\ \rho_1^0: \quad & Ax \qquad\qquad = \ b \\
\rho^1: \quad & -\bar{G}^1 x \;+\; \theta \;\geq\; \bar{g}^1 + \epsilon^1 \\
& \quad\vdots \qquad\qquad\quad \vdots \\
\rho^L: \quad & -\bar{G}^L x \;+\; \theta \;\geq\; \bar{g}^L + \epsilon^L \\
& \qquad\quad x, \quad \theta \;\geq\; 0,
\end{aligned}
\tag{2.73}
$$

where L optimality cuts have been added to the originally relaxed master problem. We do not consider feasibility cuts in the following argument, as they are exact. The vector ρ^0 and the scalars $\rho^l, l = 1, \ldots, L$, denote the dual prices. The right-hand sides $\bar{g}^l + \epsilon^l, l = 1, \ldots, L$, are independent stochastic parameters, normally distributed. We assume independence because the cuts are generated from independent samples, and we neglect the dependency due to $\hat{x}^l, l = 1, \ldots, L$, via the Benders decomposition algorithm because we assume it to be weak.

With the random parameters $\epsilon^l, l = 1, \ldots, L$, in the right-hand side, the optimum solution $\tilde{z}_M^*$ will be random. One could experimentally obtain the distribution of $\tilde{z}_M^{*L}$ by randomly varying the right-hand sides according to N samples $j = 1, \ldots, N$ drawn from the normal distributions of $\epsilon^l, l = 1, \ldots, L$, and by solving the master problem for all N samples. One could estimate the mean and the variance of the distribution from the samples $j = 1, \ldots, N$. As this is a very expensive way to obtain an estimate of the lower bound distribution, we proceed instead in the following way. We have already stated that we choose a sample size $|S|$ such that the variances $\sigma_{\tilde{z}}^{2l}, l = 1, \ldots, L$, are small. In this case we can then perform a local error analysis and assume that for all outcomes of the random right-hand sides $\epsilon^l, l = 1, \ldots, L$, the optimum solution of the master problem has the same basis. Then we can define the optimum solution of the master problem

$$\tilde{z}_M^{*L} := N(\bar{z}_M^L, \mathrm{var}(\tilde{z}_M^{*L})) \tag{2.74}$$

to be a random parameter, normally distributed with mean $\bar{z}_M^L$ and variance $\mathrm{var}(\tilde{z}_M^{*L})$, and compute the mean of the lower bound estimate by substituting the means $\bar{\epsilon}^l = 0, \ l = 1, \ldots, L$, for the random right-hand sides:

$$
\begin{array}{rlccccc}
\bar{z}_M^L = & \min\ z_M^L = & & cx & + & \theta & \\
& \text{s/t } \rho_1^0: & & Ax & & & = & b \\
& \rho^1: & & -\bar{G}^1 x & + & \theta & \geq & \bar{g}^1 \\
& \ \ \vdots & & & & & \vdots & \\
& \rho^L: & & -\bar{G}^L x & + & \theta & \geq & \bar{g}^L \\
& & & x, & & \theta & \geq & 0.
\end{array}
\tag{2.75}
$$

We compute the variance $\text{var}(\tilde{z}_M^*)$ by using the dual formulation of the master problem:

$$\text{var}(\tilde{z}_M^*) = \sum_{l=1}^{L} (\rho^l)^2 \sigma_{\tilde{z}}^2(\hat{x}^l). \qquad (2.76)$$

As the lower bound means increase monotonically with the number of iterations, we obtain the largest lower bound by

$$\tilde{\text{LB}}^L = \tilde{z}_M^{*L}, \qquad (2.77)$$

which is normally distributed:

$$\tilde{\text{LB}}^L := N(\tilde{\text{LB}}^L, \text{var}(\tilde{\text{LB}}^L)). \qquad (2.78)$$

The local error analysis can be easily extended to the case where we consider the error terms $\epsilon^l(x)$ as a function of the decision variables x instead of constant. As before, we assume the variances $\sigma_{\tilde{L}}^2(x)$ to be small so that for all outcomes of the random right-hand sides $\epsilon^l(x)$, $l = 1, \ldots, L$, the solution of the master problem has the same basis. We obtain the mean of the lower bound estimate by solving the master problem (2.75), with the means $\bar{\epsilon}^l(x^L) = 0$, $l = 1, \ldots, L$, replacing the random terms in the right-hand side. Assuming the variance of the error terms $\epsilon^l(x^L)$ to be approximately constant in a small neighborhood of x^L, we obtain the variance $\text{var}(\tilde{z}_M^*)$ by using the dual formulation of the master problem:

$$\text{var}(\tilde{z}_M^*) = \sum_{l=1}^{L} (\rho^l)^2 \sigma_{\tilde{L}}^2(\hat{x}^L). \qquad (2.79)$$

2.4.4 Stopping Rule

By analogy to the deterministic Benders decomposition algorithm, we stop if the upper and lower bound estimates are sufficiently close. In the case of probabilistic bounds, the algorithm has to be stopped if the upper and lower bounds are indistinguishable in distribution. Using a Student's t-test, we determine if $s^l > 0$ with 95% probability, where

$$s^l = \bar{\text{UB}}^l - \bar{\text{LB}}^l - \text{TOL} \tag{2.80}$$

and $\text{TOL} \geq 0$ is a given tolerance.

The employment of the Student's t-test requires independence of the upper and lower bound distributions. As independence is not ensured as an upper bound, and a binding cut in the master problem could be obtained from the same set of samples, we obtain independence by re-sampling the lowest upper bound. The $\hat{x}$ corresponding to the lowest upper bound and the corresponding importance distribution have to be stored. In each iteration we check if upper and lower bounds are close to each other, using the Student's t-test without fulfilling the independence requirement. If the bounds are close, we use new samples to compute an independent upper bound. Now we reconfirm that $s^l > 0$ by the t-test, compute a confidence interval, and quit. If the confidence interval turns out to be unsatisfactory, we must repeat the procedure with a larger sample size.

2.4.5   Confidence Interval

After passing the Student's t-test in the last iteration, which implies that the upper and lower bound means are indistinguishable, we obtain the optimum solution $\hat{x}^L$, $\hat{\theta}$ from the master problem. We derive from the distributions $\tilde{\text{LB}}^L$ and $\tilde{\text{UB}}^L$ a 95% confidence interval: on the left side by using the lower bound distribution and on the right side by using the upper bound distribution. We define

$$C_{\text{left}} = 1.96\sqrt{\text{var}(\tilde{\text{LB}}^L)}, \qquad C_{\text{right}} = 1.96\sqrt{\text{var}(\tilde{\text{UB}}^L)} \tag{2.81}$$

and obtain the confidence interval

$$\bar{\text{LB}} - C_{\text{left}} \leq Z^* \leq \bar{\text{UB}} + C_{\text{right}} \tag{2.82}$$

for the final solution Z^*. If

$$(C_{\text{left}} + C_{\text{right}})/\bar{\text{LB}}^L \leq C_{\text{tol}}, \tag{2.83}$$

where C_{tol} is a predefined quality criterion for the confidence interval, the present solution is satisfactory. Otherwise, the sample size has to be increased and the problem solved again.

2.4.6   Improvement of the Solution

Suppose the solution with a certain sample size is not satisfactory. Instead of starting from the beginning with an increased sample size, we want to use the information we have already collected. To do this, we look for the binding cuts in the final solution, increase the sample size, and recompute the binding cuts at the same $\hat{x}^l$ for which they were originally computed. This, of course, means that we have to store the values of $\hat{x}^l$ and the associated importance distributions or recompute the latter. The enlarged sample size leads to smaller variances of the binding cuts and eventually to a smaller confidence interval of the final solution. Berry-Esséen, see Hall (1982) [59], gives upper bounds on the rates of convergence for the central limit theorem. Solving the master problem again with the improved binding cuts will not in general result in an intersection of the lower and upper bound. Therefore, more iterations are necessary to obtain the optimal solution according to the increased sample size. This improvement procedure could be employed iteratively until a satisfactory solution is obtained. It may not be very efficient, however, and there may be better ways to do so. In general we choose a sample size such that the confidence interval obtained initially is satisfactory.

2.5   THE ALGORITHM

We can now state the algorithm as follows:

    Step 0   Initialize:
        $l = 0$, $\bar{\text{UB}}^0 = \infty$.

Step 1   Solve the relaxed master problem and obtain a lower bound:
$\bar{\text{LB}}^l = c\hat{x} + \hat{\theta}^l$.

Step 2 $l = l + 1$.
Solve subproblems and obtain an upper bound:
$\bar{\text{UB}}^l = \min\{\bar{\text{UB}}^{l-1}, c\hat{x}^l + \bar{z}(\hat{x}^l)\}$. Compute and add a cut to the master problem, using Monte Carlo (importance) sampling.

Step 3   Solve the master problem and obtain a lower bound:
$\bar{\text{LB}}^l = c\hat{x}^l + \hat{\theta}^l$.

Step 4   Compute $s = \bar{\text{UB}}^l - \bar{\text{LB}}^l + \text{TOL}$.
If $s > 0$ (Student's t-test), go to Step 2.

Step 5   Compute confidence interval and obtain a solution Z^*, $\hat{x}$, $\hat{\theta}$.
Stop.

Improvement of the solution:

Step 6   If $(C_{\text{left}} + C_{\text{right}})/\bar{\text{LB}} \le C_{\text{tol}}$, stop;
otherwise go to Step 7.

Step 7   Increase the sample size and initialize $\bar{\text{UB}}^0 = \infty$.

Step 8   Recompute binding cuts.
Upper bound: $\bar{\text{UB}}^l = \min\{\bar{\text{UB}}^{l-1}, C\hat{x} + \bar{z}(\hat{x}^l)\}$.

Step 9   Go to Step 3.

2.6   A CLASS OF MULTI-STAGE STOCHASTIC LINEAR PROGRAMS

Large-scale deterministic mathematical programs, used for operations and strategic planning, are often dynamic linear programs. These problems have a staircase matrix structure. Multi-stage stochastic linear programs are the stochastic extensions of these programs. In general, these programs can be extremely large because the number of scenarios grows exponentially with the number of periods. We will, however, address a certain restricted class in which the number of scenarios grows linearly with the number of stages (see Dantzig and Glynn (1990) [24] and Dantzig and Infanger (1991) [28]). The problem (whose constraints are stated below) breaks down into two parts: a deterministic dynamic (time-staged) part and a within-period stochastic part. We call the deterministic part the master problem. It is a dynamic linear program with T stages. The vectors c_t and b_t and the matrices B_{t-1} and A_t are assumed to be known with certainty.

$$
\begin{aligned}
\min \ \textstyle\sum_{t=1}^{T} c_t x_t \quad &+ \quad \textstyle\sum_{t=1}^{T} E(f_t y_t^{\omega_t}) \\
-B_{t-1}x_{t-1} + A_t x_t \quad\quad\quad\quad\quad &= \ b_t, \quad\quad t = 1,\ldots,T, \quad B_0 = 0 \\
-F_t^{\omega_t} x_t \ + \quad\quad\quad\quad D_t y_t^{\omega_t} &= \ d_t^{\omega_t}, \quad t = 1,\ldots,T, \quad \omega_t \in \Omega_t \\
x_t, \quad\quad\quad\quad\quad y_t^{\omega_t} &\geq \ 0.
\end{aligned}
$$

$$(2.84)$$

Each stage is associated with a stochastic subproblem. Uncertainty appears in the transition matrix $F_t^{\omega_t}$ and in the right-hand-side vector $d_t^{\omega_t}$, where ω_t denotes an outcome of the stochastic parameters in period t, with Ω_t denoting the set of all possible outcomes in period t. The subproblems in each stage are assumed to be stochastically independent. The subproblem costs f_t and the technology matrix D_t are assumed to be deterministic parameters.

Facility expansion planning is an example of this type of formulation. The master problem models the expansion of the facilities over time. The decision variables are the capacity built and the capacity available at time t. The subproblems model the operation of these capacities in an uncertain environment. Take, for example, the case of expansion planning of power systems. The expansion or replacement of capacities of generators and transmission lines is determined in the master problem. The capacities at each period t are made available to the system for operation. The subproblems model the power system operation, i.e., the optimal scheduling of the available capacities to meet the demand for electricity. The availabilities of generators and transmission lines and the demands are uncertain at the time when the expansion decision is made.

The approach is primarily "here and now" (Dantzig and Madansky (1961) [29]), given a context of high investment costs and long lead times for capacity expansion. However, because the operations subproblems are stochastically independent and only the expected operation costs, rather than the state of the system after period t, affect the expansion plan (as failures of equipment get repaired, and uncertainty in the demands are interpreted as deviations from a demand path), "here and now" is equivalent to "wait and see." This means that the optimal decision in period $t + 1$ depends only on the capital stock on hand at the start of period $t + 1$ and is independent of any observed outcomes in period t; i.e., the same optimal capacity expansion decision would be made before and after period t operations. Thus the facility expansion plan can be laid out at the beginning for the whole planning horizon based on the expansion costs and the expected operation costs. This permits the multi-stage problem to be treated as if it were a two-stage problem. The first "stage" concerns the single decision of what facility expansion will be scheduled in all future periods, without knowledge of the particular outcomes of the uncertain parameters in future periods. The second "stage" concerns the operations problems, where the recourse decisions made depend on the realizations of the stochastic parameters. Note that for $T = 1$, the problem is exactly a two-stage stochastic linear program with recourse. For $T \geq 2$ the problem is a two-"stage" problem with the second stage consisting of T independent subproblems. Figure 2.2 represents the decision tree corresponding to the special class of "here and now" multi-stage problems.

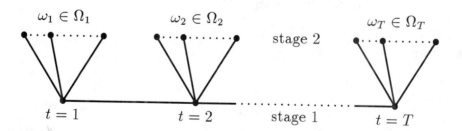

Figure 2.2: Decision tree for the special class of "here and now" multi-stage problems

Using Benders decomposition, we decompose the special class of "here and now" multi-stage problems as follows:

The master problem:

$$
\begin{aligned}
z_M^L = \min \ \textstyle\sum_{t=1}^T c_t x_t \ &+ \ \textstyle\sum_{t=1}^T \theta_t \\
-B_{t-1}x_{t-1} + A_t x_t \qquad &= \ b_t, \qquad t=1,\ldots,T, \quad B_0 = 0 \\
-G_t^l x_t \ + \qquad \alpha_t^l \theta_t \ &\geq \ g_t^l, \qquad t=1,\ldots,T, \quad l=1,\ldots,L \\
x_t \qquad\qquad\qquad &\geq \ 0,
\end{aligned}
$$

$$(2.85)$$

where the latter constraints, called cuts, are initially absent but are added in later iterations. The master problem is optimized to obtain an approximate optimal feasible solution, $x_t = \hat{x}_t^l$, that is used as input to the subproblems.

The subproblems for ω_t in period t at iteration l:

$$
\begin{aligned}
z_t^{\omega_t}(\hat{x}_t^l) \ &= \ \min f_t y_t^{\omega_t} \\
\pi_t^{\omega_t}(\hat{x}_t^l): \qquad\qquad D_t y_t^{\omega_t} \ &= \ d_t^{\omega_t} + F_t^{\omega_t}\hat{x}_t^l, \quad \omega_t \in \Omega_t, \quad t=1,\ldots,T \\
y_t^{\omega_t} \ &\geq \ 0, \qquad\qquad \hat{x}_t^l \text{ given},
\end{aligned}
$$

$$(2.86)$$

where $\pi_t^{\omega_t} = \pi_t^{\omega_t}(\hat{x}_t^l)$ are dual multipliers corresponding to the constraints and $z_t^{\omega_t} = z_t^{\omega_t}(\hat{x}_t^l)$ is the value of the objective as a function of $\hat{x}_t^l$. These are used to generate the next cut for the master problem.

The cuts:

For $t = 1, 2, \ldots, T$,

$$
G_t^l = E(\pi^{\omega_t} B^{\omega_t}), \qquad g_t^l = E(\pi^{\omega_t} d^{\omega_t}), \qquad z_t(\hat{x}_t^l) = E(z_t^{\omega_t}), \qquad \pi_t^{\omega_t} = \pi_t^{\omega_t}(\hat{x}_t^l).
$$

$$(2.87)$$

Lower (LB^l) and upper (UB^l) bounds to the problem:

$$\text{LB}^l = z_M^l, \quad \text{UB}^l = \min\{\text{UB}^{l-1}, \sum_{t=1}^{T}(c_t \hat{x}_t^l + z_t(\hat{x}_t^l))\}, \text{UB}^0 = \infty. \tag{2.88}$$

The solution procedure works analogously to the regular two-stage case. $\hat{x}_t^l$ is the optimal solution of the master problem in iteration l, and $\pi_t^{\omega_t}(\hat{x}_t^l)$ is the optimal dual solution of subproblem ω_t, given $\hat{x}_t^l$. $\alpha_t^l = 0$ corresponds to feasibility cuts and $\alpha_t^l = 1$ to optimality cuts. Solving the master problem in iteration l, we obtain a trial solution $\hat{x}_t^l$, which we pass to the subproblems. By solving sample subproblems ω_t, $\omega_t \in S_t$, in each period $t = 1, \ldots, T$, according to the importance sampling scheme, we compute estimates of the second-stage costs z_t and estimates of the gradients G_t^l and the right-hand sides g_t^l of the cuts. Note that there is one cut for each period t. The cuts are added to the master problem and the master problem is solved again. As before, the objective function value of the master problem gives a lower bound estimate, and the total expected costs of a trial solution $\hat{x}_t^l$, $t = 1, \ldots, T$, gives an upper bound estimate of the objective function value of the problem. If the lower and the upper bounds are sufficiently close according to a Student's t-test, the problem is considered to be solved. A 95% confidence interval of the optimal solution is then computed.

2.7 NUMERICAL RESULTS

The method described above has been implemented. We call the implementation the **DECIS** system, referring to **DEC**omposition and **I**mportance **S**ampling. The FORTRAN code for solving general large-scale two-stage stochastic linear problems with recourse using Benders decomposition and importance sampling uses MINOS (Murtagh and Saunders (1983) [99]), which has been adapted for this purpose as a subroutine for solving the linear programs of the master problem and the subproblems. Alternatively, the code can use a modified version of Tomlin's (1973) [120] LPM1 code of the revised simplex method as a subroutine. We have also built an interface to IBM's Optimization Subroutine Library (OSL)

(1991) [72], a collection of different linear programming, network programming, mixed-integer programming, and quadratic programming solution algorithms and data manipulation subroutines. Versions of the DECIS code are installed on several computers, including the IBM 3090 mainframe, workstations such as Digital Equipment's DECstation 5000 and IBM's RS/6000, and different personal computers.

All following test results were computed on a Toshiba T5200 laptop personal computer. First we present an illustrative example, a toy problem of expansion planning of power systems, which we discuss in detail. Then we derive numerical results from small test problems found in the literature. These problems are small enough that we are able to solve the universe case. Finally, we demonstrate the solution of large-scale test problems with numerous stochastic parameters.

2.7.1   Illustrative Example

The illustrative example, test problem APL1P, is a model of a simple power network with one demand region. There are two generators with different investment and operating costs, and the demand is given by a load duration curve with three load levels: base, medium, and peak. We index the generators with $j = 1, 2$ and the demands with $i = 1, 2, 3$. The variables x_j, $j = 1, 2$, denote the capacities that can be built and operated to meet demands d_i, $i = 1, 2, 3$. The per-unit cost to build generator j is c_j. The variable y_{ij} denotes the operating level for generator j in load level i, with operating cost f_{ij}. The variable y_{is} defines the unserved demand in load level i that must be purchased with penalty cost $f_{is} > f_{ij}$ for all j. The subscript s is not an index, but denotes only an unserved demand variable. In this way the model is formulated with complete recourse, which means that for any given choice of x, demand is "satisfied" for all outcomes.

In this model, building new generators competes with purchasing unserved demand through the cost function, yet there is a minimum capacity b_i that has to be built for each load level. The availabilities of the two generators, β_j, $j = 1, 2$, and the demands in each load level, d_i, $i = 1, 2, 3$, are uncertain. Generator 1 has four possibilities, generator 2 has five, and each demand has four. All of the data values are given in Table 2.1, and the problem can be formulated as follows:

$$
\begin{aligned}
\min \quad & \sum_{j=1}^{2} c_j x_j \;+\; E\{\sum_{j=1}^{2}\sum_{i=1}^{3} f_{ij} y_{ij}^{\omega} \;+\; \sum_{i=1}^{3} f_{is} y_{is}^{\omega}\} & & \\
\text{s/t} \quad & x_j & \geq \; & b_j & j = 1,2 \\
& -\alpha_j^{\omega} x_j \;+\; \sum_{i=1}^{3} y_{ij}^{\omega} & \leq \; & 0, & j = 1,2 \\
& \sum_{j=1}^{2} y_{ij}^{\omega} \;+\; y_{is}^{\omega} & \geq \; & d_i^{\omega}, & i = 1,2,3 \\
& x_j, \qquad\qquad\quad y_{ij}^{\omega}, \qquad y_{is}^{\omega} & \geq \; & 0, & j = 1,2, \\
& & & & i = 1,2,3.
\end{aligned}
$$

$$(2.89)$$

We will enumerate $\omega \in \Omega$ when solving the universe problem and $\omega \in S$ when solving a problem with sampling.

The number of possible demands and availabilities results in $4 \times 5 \times 4^3 = 1280$ possible outcomes in Ω. Thus 1280 subproblems have to be solved in each iteration of Benders decomposition for the universe case. We compare the universe solution with solutions gained by the importance sampling algorithm. Table 2.2 shows results for the case of 20 samples out of the possible 1280 combinations and without an improvement phase.

One hundred replications of the same experiment with different seeds were run to get statistical information about the accuracy of the solution and the estimated confidence interval. The mean over the 100 replications of the objective function value (total costs) differs from the universe solution by 0.3%. From the distribution of the optimum objective value derived from the 100 replications of the experiment, a 95% confidence interval is computed: $\pm 2.1\%$. In each replication a 95% confidence interval of the solution is estimated. The mean over all replications of the estimated confidence interval is 1.5% on the left side and 1.9% on the right side. In the worst case an objective function value of 26,233.9 was computed. This is about 6.4% off the correct answer. The estimated 95% confidence interval in this case did not cover the correct answer. The coverage rate of 90% expresses that in 90% of the 100 replications the correct answer of

Table 2.1: Model APL1P: Test Problem Data

Generator Capacity Costs (10^5 \$/(MW, a))					
$c_1 = 4.0$ $c_2 = 2.5$					

Generator Capacity Costs (10^5 \$/(MW, a))
$c_1 = 4.0$ $c_2 = 2.5$

Generator Operating Costs (10^5 \$/MW, a)
$f_{11} = 4.3$ $f_{21} = 8.7$
$f_{12} = 2.0$ $f_{22} = 4.0$
$f_{13} = 0.5$ $f_{23} = 1.0$

Unserved Demand Penalties (10^5 \$/MW, a)
$f_{1s} = f_{2s} = f_{3s} = 10.0$

Minimum Generator Capacities (MW)
$b_1 = b_2 = 1000$

Demands (MW)

#	1	2	3	4	
Outcome	900	1000	1100	1200	
Probability	0.15	0.45	0.25	0.15	

Availabilities of Generators
Generator 1 (β_1)

#	1	2	3	4	
Outcome	1.0	0.9	0.5	0.1	
Probability	0.2	0.3	0.4	0.1	

Generator 2 (β_2)

#	1	2	3	4	5
Outcome	1.0	0.9	0.7	0.1	0.0
Probability	0.1	0.2	0.5	0.1	0.1

Table 2.2: Model APL1P, 20 Samples (100 Replications of the Experiment)

	Correct	Mean	95% Conf. %	Bias %
#univ.	1280			
#iter.		7.6		
G1	1800.0	1666.5	57.0	−7.4
G2	1571.4	1732.5	52.5	10.2
Theta	13513.7	13729.4	21.3	1.6
Obj.	24642.3	24726.7	2.1	0.3
Est. conf. (%)	left	1.5		
Est. conf. (%)	right	1.9		
Coverage		0.90		

the universe solution is covered by the estimated confidence interval. This shows that when we use a sample size of 20, we are slightly underestimating the confidence interval: if the computation of the 95% confidence interval were exact, we would expect a coverage rate of 95%. The reason for the underestimation of the 95% confidence interval lies in the underlying assumptions of the estimation method, e.g., a normal error distribution for sample sizes of 20, constant error distribution along a cut, and the same basis for all outcomes of the random right-hand sides of the cuts. The latter assumption especially is true only if the variances are small. A larger sample size reduces the variances, and we expect a better coverage rate for the 95% confidence interval. The bias and the confidence interval of the optimum strategies (the loads x to be installed) are larger than those of the optimum objective function value. The objective function near the optimal solution appears to be flat: several different strategies yield close to the optimum costs. Confidence intervals for the two components of x of about 57% and 52% are computed. In the above example a sample size of 20 was chosen. Note that additional computational effort is also needed to obtain the importance distribution; e.g., 17 subproblems have to be solved in each iteration to obtain the marginal costs M_i. Compared to the universe solution, the method achieves with about 2.9% the computational effort a solution that is with 95% confidence within an interval of ±2.1% of the correct answer. We conclude that importance sampling seems to be a promising approach to solving stochastic linear programs.

Table 2.3: Model APL1P, 200 Samples (100 Replications of the Experiment)

	Correct	Mean	95% Conf. %	Bias %
#univ.	1280			
#iter.		7.9		
G1	1800.0	1728.7	31.5	−4.0
G2	1571.4	1681.7	29.2	7.0
Theta	13513.7	13554.7	12.2	0.3
Obj.	24642.3	24673.8	0.4	0.1
Est. conf. (%)	left	0.4		
Est. conf. (%)	right	0.7		
Coverage		0.95		

Table 2.3 represents the results when using 200 samples. It shows decreasing bias, decreasing confidence intervals, and improving estimations of the confidence intervals with increased sample size. The coverage of the 95% confidence interval, computed by 100 replications of the experiment with different seeds, is now 95%.

2.7.2 Test Problems from the Literature

We investigate the performance of the algorithm on two other examples from the literature that are small enough to compute the universe solution. PGP2, derived from Louveaux and Smeers (1988) [89], is a power generation–planning model used to determine the capacities of various types of equipment required to ensure that consumer demand is met. The demands in three demand regions are stochastic and represented by discrete random variables with 9, 9, and 8 outcomes. CEP1 is a capacity-planning model for a manufacturing plant in which several parts are produced on several machines. If the demand for the parts exceeds the production capability, the residual parts are purchased from external sources at a price much higher than the production costs to meet the demand. There are three stochastic parameters (demands for parts), with discrete uniform distributions covering 10 outcomes each. The formulations and data for CEP1 and PGP2 may be found in Higle, Sen, and Yakowitz (1990) [63]. Table 2.4 and Table 2.5 represent the computational results of PGP2 and CEP1 and show the size of the test problems.

In the case of PGP2 we obtained very accurate results using a sample size of 50. By computing 100 replications of the experiment, we found that the mean of the objective function values differs 0.1% from the correct answer. The 95% confidence interval of the objective function value, computed by the 100 replications of the experiment, is ±0.76%. The mean of the confidence intervals estimated in each replication is 0.62% on the left side and 0.9% on the right side. In 98% of the cases the correct solution is covered by the 95% confidence interval. In the worst case the solution differed by 0.77% from the correct answer and was not covered by the 95% confidence interval.

In the case of CEP1 a higher sample size is needed to obtain accurate results. The estimation of the second-stage costs appears to be more difficult. The reason lies in the fact that the (penalty) costs of buying parts from external sources are much higher than the costs of production. For this problem the additive approx-

Table 2.4: Model PGP2, 50 Samples (100 Replications of the Experiment)

		Correct	Mean	95% Conf. %	Bias %
#univ.		648			
#iter.			9.1		
Obj.		392.2	392.5	0.76	0.1
Est. conf. (%)	left		0.62		
Est. conf. (%)	right		0.9		
Coverage			0.98		
Comp. time (min)			0.28		
Problem Size					
Master:	rows		3		
	columns		7		
	nonzeros		16		
Sub:	rows		8		
	columns		16		
	nonzeros		52		

Table 2.5: Model CEP1, 200 Samples (100 Replications of the Experiment)

		Correct	Mean	95% Conf. %	Bias %
#univ.		1000			
#iter.			6.4		
Obj.		57790.7	58832.7	4.63	1.8
Est. conf. (%)	left		4.65		
Est. conf. (%)	right		4.62		
Coverage			0.95		
Comp. time (min)			0.28		
Problem Size					
Master:	rows		12		
	columns		10		
	nonzeros		36		
Sub:	rows		9		
	columns		16		
	nonzeros		53		

imation function is not a very good approximation to the true cost function, as it does not cover the very high costs in scenarios where all three demands are high. The estimated confidence interval seems to be large. We computed 4.65% on the left side and 4.62% on the right side (mean over 100 replications of the experiment). The estimations of the confidence interval are accurate as indicated by the coverage rate of 95% of the correct answer by the 95% confidence interval. In the worst case a difference of 8.07% of the objective function value to the correct answer was computed. The worst-case solution is not covered by the estimated confidence interval. In this example it is easier to compute the value of the first-stage variables than to estimate the second-stage costs. In most cases the correct answer for the first-stage variables was obtained. We have developed methods that adaptively improve the approximation function if sample information shows that the variance of the estimation is too large.

2.7.3   Large-Scale Test Problems

We next report on the solution of large-scale test problems with several stochastic parameters. These problems are so large that it is impossible to compute the universe solution.

Facility Expansion Planning

WRPM is a prototype multiarea capacity-expansion-planning problem for the western USA and Canada. The model is detailed, covering six regions, three demand blocks, two seasons, and several kinds of generation and transmission technologies. The objective is to determine optimum discounted least-cost levels of generation and transmission facilities for each region of the system over time. The model minimizes costs of supplying electricity (investment and operating costs) to meet the exogenously given regional demand subject to expansion and operating constraints. A description of the model can be found in Dantzig et al. (1989) [25] and Avriel, Dantzig, and Glynn (1989) [3]. In the stochastic version of the model, the availabilities of generators and transmission lines and the demands are subject to uncertainty. There are 13 stochastic parameters per time period (8 stochastic availabilities of generators and transmission lines and 5 uncertain demands) with discrete distributions over three or four outcomes. The operating subproblems in each period are stochastically independent. WRPM has been developed in three flavors. The test problem WRPM1 covers a time horizon of one future period, and WRPM2 covers two future periods of 10 years each. WRPM1 and WRPM2 are reduced versions of WRPM3, the largest problem with the most realistic formulation. There are differences in the parameters for WRPM1, WRPM2, and WRPM3. Note that, in the deterministic equivalent formulation, the problem would have more than 1.5 billion (WRPM1) and more than 3 billion (WRPM2) equations.

Computational results for these large-scale test problems are represented in Table 2.6. Besides the solution of the stochastic problems, Table 2.6 shows the results from solving the expected-value problem. In this case the stochastic parameters are replaced by their expectations to obtain a deterministic problem. The expected-value solution is then used as a starting point for the stochastic solution. We also report on the estimated expected costs of the expected-value solution. These are the total expected costs that would occur if the expected-value solution were implemented in a stochastic environment. The objective

Table 2.6: Large Test Problems: Computational Results for Power Planning

		WRPM1	WRPM2	WRPM3
# iter. stoch. (exp. val.)		139 (82)	131 (83)	197 (129)
Sample size		100	100	100
Exp.-val. solution obj.		286323.2	140041.0	196471.4
Exp.-val. solution, exp. cost		295473.7	147227.3	202590.3
Stochastic solution		289644.2	143109.2	199017.4
Est.	conf. left %	0.0913	0.0962	0.0292
	conf. right %	0.063	0.1212	0.067
	solution time (min)	75	187	687
Problem Size				
Master	rows	44	86	128
	columns	76	151	226
	nonzeros	153	334	413
Sub	rows	302	302	302
	columns	289	289	289
	nonzeros	866	866	866
# stochastic parameters		13	26	39
# universe scenarios		5038848	10077696	$15 \cdot 10^6$

value for the true stochastic solution must lie between the objective value of the expected-value solution and the expected costs of the expected-value solution.

In all cases (WRPM1, WRPM2, and WRPM3), we chose a sample size of 100. The estimates of the objective function value of the stochastic solution (289,644.2 in case of WRPM1 and 143,109.2 in case of WRPM2) are amazingly accurate. The 95% confidence interval was computed as 0.0913% on the left side and 0.063% on the right side (WRPM1) and 0.0962% on the left side and 0.1212% on the right side (WRPM2). Thus the objective value of the stochastic solution lies with 95% probability within $289,379.7 \leq z^* \leq 289,826.0$ (WRPM1) and $142,971.5 \leq z^* \leq 143,282.6$ (WRPM2). In both cases the expected costs of the expected-value solution and the expected costs of the stochastic solution differ significantly. The solution times on a Toshiba T5200 laptop PC with 80387 mathematic coprocessor were 75 minutes (WRPM1) and 187 minutes (WRPM2). During this time about 7500 (WRPM1) and 15,700 (WRPM2) subproblems (linear programs of the size of 302 rows and 289 columns) were solved.

The realistic model WRPM3 covers a time horizon of three future periods of 10 years each. Thus the total number of stochastic parameters is 39. The number of universe scenarios is larger than $5 \cdot 10^6$ per period. The deterministic equivalent formulation of the problem, if it were possible to state it, would have more than 4.5 billion constraints.

The stochastic WRPM3 was solved by using a sample size of 100 per period. It took 129 iterations to obtain the expected-value solution and an additional 68 iterations to compute the stochastic solution. The objective function value of the stochastic solution was estimated as 199,017.4 with a remarkably small 95% confidence interval of 0.029% on the left side and 0.067% on the right side. Thus the optimal solution lies with 95% confidence between $198,959.3 \leq z^* \leq 199,164.1$. The expected costs of the expected-value solution (202,590.3) and the objective function value of the stochastic solution differ significantly. The problem was solved in 687 minutes on the Toshiba T5200. This includes time to solve 26,295 linear subproblems with 302 rows and 289 columns, and 197 master problems.

Note that the solution times of WRPM1, WRPM2, and WRPM3 do not increase linearly with the size of the system. WRPM3 takes significantly longer to solve than WRPM1 and WRPM2. We have used LPM1 to solve WRPM1 and WRPM2 and MINOS for solving WRPM3. Although the LPM1 code is less sophisticated and less reliable than MINOS, it solves a series of subproblems, where we start each optimization from the optimal solution of the previous subproblem, faster than MINOS. This is primarily because LPM1 has less computational overhead than MINOS. In addition to the difference in the code used for solving the problems, WRPM3 takes more iterations than WRPM1 and WRPM2, which results in more computational work to be performed and in a larger solution time.

Portfolio Management

Computational results from portfolio management problems can be found in Table 2.7. FI12 is an example, formulated as a network problem. It is a modified version of test problems found in Mulvey and Vladimirou (1989) [96]. The problem is to select a portfolio that maximizes expected returns in future periods, taking into account the possibility of revising the portfolio in each period. There are also transaction costs and bounds on the holdings and turnovers. The test problem FI12 covers a planning horizon of two future periods. The returns of

Table 2.7: Large Test Problems: Computational Results for Financial Planning

		FI12	LP42
# iter. stoch. (exp. val.)		4 (2)	4 (6)
Sample size		200	600
Exp.-val. solution obj.		1.0766	1.611
Exp.-val. solution, exp. cost		1.172	2.334
Stochastic solution, obj.		1.169	2.329
Est.	conf. left %	0.454	0.536
	conf. right %	0.371	0.767
	solution time (min)	2	209
Problem Size			
Master	rows	48	49
	columns	33	83
	nonzeros	130	133
Sub	rows	61	178
	columns	45	309
	nonzeros	194	570
# stochastic parameters		26	52
# universe scenarios		$2.5 \cdot 10^{12}$	$6 \cdot 10^{24}$

the stocks in the two future periods are stochastic parameters. The problem is formulated as a two-stage problem. Rather than solving the problem by looking at a certain number of preselected scenarios (18 to 72 in the case of Mulvey and Vladimirou), we assumed the returns of the stocks in the future periods to be independent random parameters, discretely distributed with three outcomes each. Because there are 13 stocks with uncertain returns, the problem has 26 stochastic parameters. The universe number of scenarios $(2.5 \cdot 10^{12})$ is very large, so the deterministic equivalent formulation of the problem has more than 10^{14} rows. The stochastic parameters appear in the B matrix as well as in the D matrix.

A sample size of 200 was chosen for solving test problem FI12. The problem was solved in only four iterations. The objective function value of the stochastic solution is computed as 1.1695, with a 95% confidence interval of 0.454% on the left side and 0.371% on the right side. Thus with 95% probability the optimal solution lies in the range $1.164 \le z^* \le 1.174$. The estimated expected costs of the expected-value solution (1.172) lie within the 95% confidence interval of the costs of the stochastic solution. However, in this case the expected costs of the expected-value solution and the expected costs of the stochastic solution differ significantly.

LP42 is also a portfolio management test problem, formulated as a network problem. It is a modified version of test problems found in Mulvey and Vladimirou (1989) [96] and of the same structure as FI12 but much larger. Again, there are transaction costs and bounds on the holdings and turnovers. The test problem covers a planning horizon of four future periods. The returns of the stocks in the four future periods are assumed to be independent stochastic parameters, discretely distributed with three outcomes each; this formulation differs from that of Mulvey and Vladimirou, who restricted the problem size by looking at a certain number of preselected scenarios. As in Mulvey and Vladimirou the multiperiod problem is viewed as a two-stage problem, where all future periods are included in the second stage. With 13 stocks with uncertain returns, the problem has 52 stochastic parameters. The universe number of scenarios $6 \cdot 10^{24}$ is very large, so the deterministic equivalent formulation of the problem, if expressed explicitly, would have more than $1.9 \cdot 10^{27}$ rows. Here, the stochastic parameters appear in the B matrix as well as in the D matrix. Because of the stochasticity of the D matrix, cuts from the expected-value problem are not valid for the stochastic problem. The expected-value problem and the stochastic problem were solved separately. A sample size of 600 was chosen, and the solution (objective function value 2.329) was obtained in four iterations. The 95% confidence interval is very small given the large number of stochastic parameters, namely 0.536% on the left side and 0.767% on the right side. Thus with 95% confidence the objective value of the optimal solution lies within $2.316 \leq z^* \leq 2.347$. The expected costs of the expected-value solution are significantly different from the expected costs of the stochastic solution.

3

Using Parallel Processors

In collaboration with James K. Ho we have explored how our approach for solving two-stage stochastic linear programs can be effectively implemented on a parallel (Hypercube) multicomputer (Dantzig, Ho, and Infanger (1991) [26]).

3.1 HYPERCUBE MULTICOMPUTERS

Advances in VLSI (very large-scale integration) for digital circuit design are leading to much less expensive and much smaller computers. They have also made it possible to build a variety of "supercomputers" consisting of many small computers combined into an array of concurrent processors. We shall refer to such an architecture as multicomputers. Each individual processor is called a node. Typically, the nodes are the same kind as those used in high-end microcomputers and are relatively inexpensive. Significant computational power can be obtained by making many of them work in parallel at costs that are much lower than an equivalent single processor. Obviously, the effectiveness of the approach depends on whether an application can be reduced to a well-balanced distribution of asynchronous tasks on the nodes. Linear programming and especially stochastic linear programs solved by decomposition naturally fit into this framework.

A Hypercube multicomputer is essentially a network of 2^n processors interconnected in a binary n-cube (or hypercube) topology. The connections for $n \leq 4$ are illustrated in Figure 3.1. The figure demonstrates the evolution of nodes and

n=0

n=1

n=2

n=3

n=4

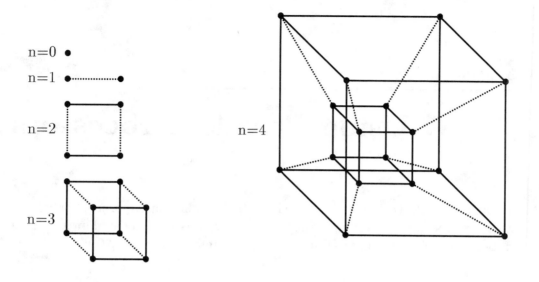

Figure 3.1: Hypercube multicomputers of dimensions $n \leq 4$

connections when the dimensionality is increased from dimension $n-1$ to dimension n: a copy of the $n-1$ dimensional network is put beside its original, and each pair of corresponding nodes is interlinked by a new connection. New connections are represented by dotted lines; solid lines represent the original connections of the $n-1$ dimensional network.

Each processor (or node) has its local memory and runs asynchronously of the others. Communication is by means of messages. A node can communicate directly with its n neighbors. Messages to more distant nodes are routed through intermediate nodes. The hypercube topology provides an efficient balance between the costs of connection and the benefits of direct linkages. Usually, a host computer serves as an administrative console and as a gateway to the hypercube for users.

For the work reported here, we used an Intel iPSC/2 d6 with 64 nodes at the Oak Ridge National Laboratory. Each node consists of Intel's 32-bit 80386 CPU (4 MIPS) coupled with an 80387 (300 Kflops) numeric coprocessor for floating-point acceleration. It has 4 MBytes of local memory. The hypercube (or Cube) is accessed via a host (or System Resource Manager), which is also an 80386-based system with 8 MBytes of memory and a 140-MByte hard disk. The operating system on the host is Unix System V/386 (Release 3.0). The data transfer rate between the System Resource Manager and the Cube has a peak value of 2800 KBytes/sec.

Although the nodes are physically connected as the edges of a hypercube, a trademarked routing network called DIRECT-CONNECT provides essentially uniform communication linkages between all the nodes. The hypercube can therefore be programmed as an ensemble of processors with an arbitrary communications network in which each node can communicate more or less uniformly with all other nodes. The host machine allows the user to perform the following tasks:

- Edit, compile, and link host/node programs.

- Access and release the cube (or a partition thereof).

- Execute the host program.

- Start or kill processes on the cube.

Operations peculiar to the hypercube are controlled either by Unix-type commands (iPSC/2 commands) or by extensions to standard programming languages such as Fortran and C (iPSC/2 routines). The iPSC/2 commands and routines for the Fortran programming environment are documented in Intel (1988) [75] and [76]. To execute a typical parallel program, the following steps are used:

- Compile and link the host and node programs to create executable modules.

- Obtain a partition of the cube (a sub-cube) of suitable size. For example, a call for a sub-cube of dimension 3 allocates to the user an exclusive sub-cube of eight processors identified by the node numbers $0, 1, 2, \ldots, 7$.

- Run the host program. Node programs are loaded onto the appropriate nodes at runtime in response to calls in the host program.

- On termination, kill all node processes and flush messages.

- Relinquish access to the sub-cube.

Internodal communication and host-to-node communication is done with sub-routine calls in the corresponding programs.

3.2   THE PARALLEL ALGORITHM

The Hypercube computer has an architecture of loosely coupled multiprocessors. The nodes of the cube are independent processors, where each processor has its own operating system and its own memory. The nodes are connected via a communication network. Information is exchanged between nodes only by sending messages. The hypercube architecture defines which nodes are directly connected and which nodes are only indirectly connected via third nodes. Message-routing systems of modern Hypercube computers, such as the Intel iPSC/2 computer that we are using, ensure that communication between indirectly connected nodes is very fast. Thus the difference in the communication time between directly and indirectly connected nodes is negligible. However, the time spent for communication can be significant if much information is exchanged between nodes. Therefore, the design of a parallel algorithm for loosely connected multiprocessors should be laid out in such a way that only minimum amounts of information have to be exchanged between nodes.

The main work is in the repeated solving of the master problem, and in the solving of the subproblems in the preparatory phase and in the sample phase. All other tasks are comparatively unimportant with respect to computing time. We assign processor 0 to be the master processor. Besides its main task of solving the master problem, the master processor also controls the computation and synchronizes the algorithm. The other processors (1–63) are assigned to be subprocessors, with the main task of solving subproblems. This design requires communication between the master processor and the subprocessors. No information needs to be exchanged between different subprocessors.

In addition, there is a host processor that has access to data storage devices and manages data input and output. The execution of the parallel program follows the following general steps. The host processor loads the host module (the executable file for the host processor) into its memory and starts the execution. Next, the executable files for the master processor and the subprocessors are loaded into the host and then sent to the master processor and the subprocessors, respectively. The master processor and the subprocessors start execution after they receive their modules. After processing the input data and sending it to master and subs, the host remains inactive until it receives the optimal solution from the master processor. During this time the algorithm is performed entirely in the cube, and the master processor controls the execution of the program. After receiving the optimal solution, the host processor outputs the solution to the disk, stops the execution of the programs of the master and subprocessors, and releases the cube, terminating the parallel program.

The problem data include the problem specification of the master and the sub, the stochastic information, and control parameters for the execution of the program. The input data for specifying the master problem and the subproblem are given in the form of an MPS file. Internally the problems are stored in the form of the data structures used by the linear programming solver, which we use as a subroutine. We adapted LPM1 (Tomlin 1973) [120], a linear programming optimizer, for our purposes. Clearly, the master processor receives only the data for the master problem and the subprocessors get only the data for the subproblem. Thus no switching between different problems is necessary, as it would be in a serial implementation. Both master processor and subprocessor receive the complete stochastic information. The stochastic data include the identification of the stochastic parameters within the problem and their distributions.

An index vector $\nu^\omega = (\nu_1, \ldots, \nu_h)^\omega$ completely defines a scenario ω. We define $\nu_i \in \Omega_i$ or $\nu_i \in \{1, \ldots, k_i\}$, $i = 1, \ldots, h$. For example, $\nu^\omega = (1, 3, 2)$ would denote a scenario given by the first outcome of random parameter 1, the third outcome of random parameter 2, and the second outcome of random parameter 3. Thus only the index vector ν^ω is transmitted from the master processor to a subprocessor to identify the scenario subproblem to be solved. For example, for $h = 20$ and a 4-byte integer representation, 80 bytes have to be sent. Besides the scenario information ν^ω, the current solution of the master problem $\hat{x}^l$ is needed to set up the scenario problem ω. We pass $\hat{x}^l$ to each subprocessor only once per iteration, at the beginning of the preparation phase. The flag $I_x \in \{0, 1\}$ tells the subprocessor if an $\hat{x}$ has to be received (1) or not (0).

Now each subprocessor looks up the outcomes of the stochastic parameters corresponding to its ν to set up the vector $b(\nu)$ and the matrix $B(\nu)$. Using $\hat{x}$, the right-hand side $b(\nu) + B(\nu)\hat{x}$ is computed and the subproblem is solved.

In all cases the optimal objective function value $z(\nu)$ has to be sent to the master processor. Dual information for the coefficients G and the right-hand side g of the cut is all that is needed from the base-case scenario and all sample scenarios. In this case we compute the products $G(\nu) = \pi(\nu)B(\nu)$ and $g(\nu) = \pi(\nu)b(\nu)$ and send the result to the master processor. The flag I_c tells a subprocessor whether the computation and the sending of $G(\nu)$ and $g(\nu)$ is requested (1) or not (0).

In our design the subprocessors do not have any information as to the status of the algorithm. The subprocessors set up and solve the subproblems and post-process the solution. The computation is controlled by the master processor through the flags I_x and I_c.

The master processor runs the entire algorithm except for obtaining solutions of subproblems. An important task is controlling the assignment of subproblems ω to subprocessors j in the case where more sample subproblems have to be solved per iteration than there are subprocessors available. Assigning subproblems in equal proportions to subprocessors is not always possible for all sample sizes, nor is it most efficient. Different subproblems need different amounts of time to be solved. The solution time mainly depends on how many columns of the starting basis (from which the solving procedure is started) differ from the optimal basis of the subproblem. Clearly, it makes sense and is convenient to use as a starting basis the optimal basis of the subproblem that was last solved on the same processor.

We implemented an algorithm to adaptively balance the workload of the subprocessors. In our scheme the master processor keeps track of whether a subprocessor j is busy or idle. At the beginning of each solving phase (preparatory and sample phase), all subprocessors are idle. The master processor starts a subprocessor working by sending the first message (I_x, I_c) to it. At this time the subprocessor is set to busy. It is set to idle again when its solution has arrived at the master processor. Given a queue of subproblems to be solved, the first subproblem in the queue is assigned to the next idle subprocessor. The master processor keeps switching between sending out problems and receiving solutions until all subproblems are solved. Of course, the mapping $\omega \rightarrow j$ is not unique because different subproblems ω are solved by one subprocessor j. However, because we send a new problem only after the solution of the previous problem has been received, the solution of a subproblem ω can be identified as uniquely coming from subprocessor j.

We can now summarize and state the algorithm. Step 2 (in the following, Step 2 is referred to as Step M: 2.0 till Step M: 2.6 and Step S: 2.1 till Step S: 2.6) is computationally the most expensive part and is the part performed using parallel processors.

The Algorithm

Host processor

Step H: 0.0   Load host executable module from disk.

Step H: 0.1   Load master module from disk.
Send master module to processor 0.
Load submodule from disk.
Send submodule to processors j, $j = 1, \ldots, J$.

Step H: 0.2   Read data from disk.
Send control data and stochastic data to processors j, $j = 0, \ldots, J$.
Send master problem data to processor 0.
Send subproblem data to processors j, $j = 1, \ldots, J$.

Step H: 6   Receive optimal solution.
Write solution report.
Kill (free) cube. Stop.

Master Processor

Step M: 0   Receive master module from host processor.
Receive control and stochastic data from host processor.
Receive master problem data from host processor.
Initialize: $l = 0$, $\mathrm{UB}^0 = \infty$.

Step M: 1   Solve the relaxed master problem.
Obtain a trial solution $\hat{x}^l$ and a lower bound LB^l.

Step M: 2.0 $l = l + 1$.

Step M: 2.1   Determine preparatory scenarios $\nu^\omega = (\nu_1, \ldots, \nu_h)^\omega$, $\omega = 1, \ldots, n_{\text{prep}}$ according to the importance sampling scheme.

Step M: 2.2   For $\omega \in \{1, \ldots, n_{\text{prep}}\}$:
Determine $\omega \longrightarrow j$.
Send $I_x(j)$ and $I_c(\omega)$ to subprocessor j.
Send $\hat{x}^l$ to subprocessor j.
Send ν^ω to subprocessor j.
For $\omega \in \{1, \ldots, n_{\text{prep}}\}$:
Receive z^ω from subprocessor j.
If $I_c(\omega) = 1$: Receive G^ω and g^ω from subprocessor j.

Step M: 2.3   Compute the importance distribution.

Step M: 2.4   Sample scenarios $\nu^\omega = (\nu_1, \ldots, \nu_h)^\omega, \omega = 1, \ldots, n$ from the importance distribution.

Step M: 2.5   For $\omega \in \{1, \ldots, n\}$:
Determine $\omega \longrightarrow j$.
Send $I_x(j)$ and $I_c = 1$ to subprocessor j.
Send ν^ω to subprocessor j.
For $\omega = 1, \ldots, n$:
Receive z^ω from subprocessor j.
Receive G^ω, g^ω from subprocessor j.

Step M: 2.6   Obtain estimates of the expected second-stage costs and of the coefficients and right-hand side of the cut. Add the cut to the master problem. Obtain an estimate of the upper bound UB^l.

Step M: 3   Solve the master problem. Obtain a trial solution $\hat{x}^l$ and a lower bound LB^l.

Step M: 4 $s = \text{UB}^l - \text{LB}^l - \text{TOL}$
If $s \geq 0$ (Student t-test) go to Step M: 2.0.

Step M: 5   Obtain a solution and compute a confidence interval.

Step M: 6   Send optimal solution to host processor.

Subprocessor j:

Step S: 0 Receive submodule from host processor.
Receive control and stochastic data from host processor.
Receive subproblem data from host processor.

Step S: 2.1 Receive I_x and I_c from the master processor.
If $I_x = 1$: Receive $\hat{x}$ from the master processor.
Receive ν from the master processor.

Step S: 2.2 Compute $B(\nu)$ and $b(\nu)$ and the right-hand side $b(\nu) + B(\nu)\hat{x}$.

Step S: 2.3 Solve scenario subproblem ν.

Step S: 2.4 Send $z(\nu)$ to the master processor.

Step S: 2.5 If $I_c = 1$:
Compute $G(\nu) = \pi(\nu)B(\nu)$, $g(\nu) = \pi(\nu)b(\nu)$.
Send $G(\nu)$ and $g(\nu)$ to the master processor.

Step S: 2.6 Go to Step S: 2.1.

3.3 PERFORMANCE MEASURES

The main purpose of parallel processing is to reduce elapsed computing time relative to conventional sequential computation. When large sample sizes are necessary to obtain good approximate solutions to stochastic linear programs, parallel processing is an important part of the solution technique, because the solution times on sequential computers may be excessive.

Assuming that a number p of processors are available and allocated to solve the problem at hand, we compare the parallel time utilizing p processors to the sequential time using only one processor. We define the parallel time t_p as the duration from start to finish of the solution process in the parallel implementation. In terms of CPU times, t_p covers the disjoint union (nonoverlapping total) of the CPU times of all processors. We define the sequential time t_s to be the sum of all CPU times of all processors. The sequential time t_s differs from a sequential time obtained by actually solving the problem on one processor. This would require a different implementation and would not be directly comparable. In a serial version no messages are sent. On the other hand, computing resources are needed for alternately switching between solving the master problem and solving the subproblems.

The speedup S in using p processors instead of one is given by

$$S = \frac{t_s}{t_p}. \tag{3.1}$$

The efficiency is defined by

$$E = \frac{S}{p} \times 100\%. \tag{3.2}$$

A simple set of algebraic formulas can be used to predict the sequential time t_s and the parallel time t_p. We denote by t_{MA} the mean duration to compute the tasks assigned to the master processor per iteration. We define t_{SUB} to be the mean duration to compute the tasks assigned to a subprocessor (mainly solving one subproblem) when starting from the optimal solution of the previously solved subproblem. The duration t_{SUB}^0 is the mean additional time if solving a subproblem from scratch. One can also think of it as the mean duration to obtain a "good" starting basis. Thus, with L being the number of iterations,

$$\frac{t_s}{L} = t_{\mathrm{MA}} + t_{\mathrm{SUB}}^0 + (n_{\mathrm{prep}} + n)\, t_{\mathrm{SUB}} \tag{3.3}$$

and

$$\frac{t_p}{L} = \begin{cases} t_{\mathrm{MA}} + t_{\mathrm{SUB}}^0 + \frac{(n_{\mathrm{prep}}+n)}{p-1} t_{\mathrm{SUB}}, & \text{if } n, n_{\mathrm{prep}} \geq p - 1; \\ t_{\mathrm{MA}} + t_{\mathrm{SUB}}^0 + (1 + \frac{n}{p-1}) t_{\mathrm{SUB}}, & \text{if } n \geq p - 1, n_{\mathrm{prep}} < p - 1. \end{cases} \tag{3.4}$$

If the sample size n is smaller than the number of subprocessors, the parallel algorithm is not efficient because not all computer resources are utilized. Using the above formulas, we can compute the efficiency for the case of $n, n_{\text{prep}} \geq p - 1$, for example, as

$$E = \frac{t_{\text{MA}} + t_{\text{SUB}}^0 + (n_{\text{prep}} + n)t_{\text{SUB}}}{p\, t_{\text{MA}} + p\, t_{\text{SUB}}^0 + \frac{p}{p-1}(n_{\text{prep}} + n)t_{\text{SUB}}}. \tag{3.5}$$

One can see that, for a fixed number of processors, the efficiency approaches 100% as the sample size increases. This is obvious because increasing the sample size means adding computational work that can be conducted in parallel. Thus the parallel implementation is most efficient when solving problems with large sample sizes. On the other hand, one can also see that, for a given sample size, the efficiency decreases with an increasing number of processors. The maximum number of processors that can be utilized meaningfully is $1 + \max\{n_{\text{prep}}, n\}$.

3.4 NUMERICAL RESULTS FOR THE PARALLEL IMPLEMENTATION

Experiments were conducted to validate the parallel implementation and to obtain measures of computing time, speedup, and efficiency. Test problems taken from the literature are usually small, with a small number of stochastic parameters. We tested our parallel-processing methodology on a truly large-scale problem, BIGNEW, which is a modified version of the capacity-expansion-planning model WRPM, as discussed in Section 2.7.3. There are 11 stochastic parameters (8 stochastic availabilities of generators and transmission lines and 3 uncertain demands) with discrete distributions with three or four outcomes. Although other implementations of WRPM cover up to three future time periods, BIGNEW covers a planning horizon of only one future time period and is formulated as a two-stage stochastic linear program with recourse. The problem is large-scale, though it is far from the largest we have solved serially. The number of universe scenarios is about 10^6; the equivalent deterministic formulation of the problem (if it were possible to state it explicitly) would have more than 0.3 billion constraints.

This test problem has been solved repeatedly using different numbers of processors and different sample sizes. The parallel implementation has been improved as we learned more about its characteristics. For example, we varied the sample size within the range 20 to 63, where we always have at least as many processors at hand as there are subproblems to be solved in one parallel phase. Table 3.1 summarizes the results. The computing time (measured in CPU minutes per iteration) is approximately constant at a level of 0.12 minutes per iteration for sample size 20 up to 29. Then it jumps to a level of approximately 0.17 minutes per iteration, where it again remains approximately constant.

In the test example the number of preparatory subproblems to compute the importance distribution is 29. Figure 3.2 shows how the algorithm parallelizes. It schematically shows busy and idle times for different processors with sample size 63 during the first two iterations. Note the two phases of solving subproblems: the preparatory phase and the sample phase. In the preparatory phase only 29 subproblems have to be solved, compared to 63 subproblems in the sample phase. Each optimization is started using the basis for the optimal solution of the problem previously solved on the same processor. At the beginning, all problems are started from scratch because no basis is available. In the first iteration processors 1 to 29 start from scratch in the preparatory phase, but they use the optimal bases from the preparatory subproblems in the sample phase.

Table 3.1: Effect of Warm Starting All Subs

Nodes reser.	p	n	Iter.	No Warm start		Warm start		
				CPU (min)	Time/it.	CPU (min)	Time/it.	Obj.
32	32	20	66	7.898	0.120	7.898	0.120	188382
32	32	24	63	7.502	0.119	7.502	0.119	188025
32	32	26	61	7.866	0.129	7.866	0.129	188236
32	32	27	56	6.219	0.111	6.319	0.111	188232
32	32	28	52	6.434	0.124	6.434	0.124	188195
32	32	29	60	7.303	0.122	7.303	0.122	188492
32	32	30	64	11.173	0.175	7.770	0.121	188271
32	32	31	60	10.767	0.179	7.331	0.122	188301
64	33	32	64	12.409	0.194	N/A	N/A	188347
64	36	35	59	10.334	0.175	N/A	N/A	188295
64	41	40	63	10.898	0.173	7.516	0.119	188261
64	51	50	63	11.034	0.175	7.528	0.119	188378
64	61	60	70	12.035	0.172	8.374	0.120	188549
64	64	63	75	12.645	0.169	8.821	0.118	188492

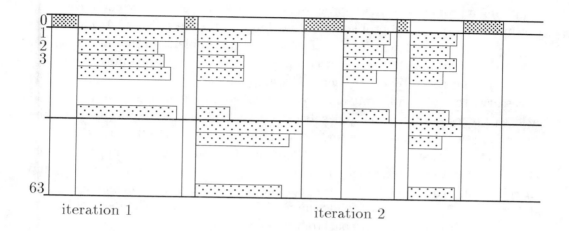

iteration 1 iteration 2

Figure 3.2: The parallel algorithm

Processors 30 to 63 do not solve subproblems in the preparatory phase; thus the sample subproblems assigned to these processors are started from scratch.

Solving a subproblem from scratch takes considerably more time than solving it with a good starting basis (warm start). The master processor starts operation when all necessary subproblems are solved completely, both in the preparatory phase and the sampling phase. The computing time in each phase is determined by the maximum duration spent solving a subproblem. In the first iteration processors 30 to 63 are idle during the preparation phase and solve subproblems from scratch in the sample phase; the maximum time spent in the sample phase by these processors is much larger than the maximum time spent by processors 1 to 29. The duration of the sample phase in the first iteration is therefore much larger for sample sizes larger than 29, the number of preparatory subproblems. The jump in the computing time at sample size 30 is due to this effect.

Besides the impact of the starting basis in the first iteration, there is also an impact in all other iterations. A basis from the optimal solution of a subproblem of the current iteration is expected to be a better starting basis than a basis from the optimal solution of a subproblem of the previous iteration. Note that the effect occurs only if $n_{\mathrm{prep}} < p - 1$ and $n > n_{\mathrm{prep}}$. We overcome this effect by supplying a proper basis to subprocessors 30 to 63. In general, one could copy the optimal basis of the subproblem that has finished first in the preparatory phase to processors 30 to 63 in order to warm-start all subproblems in the sample phase. Because idle processors are not used for any other tasks and cannot be used in a timesharing mode by other users, it is more efficient (as no communication is necessary) to assign an arbitrary preparatory subproblem (e.g. subproblem 1) also to processors 30 to 63 and solve it on each of the processors to have the optimum starting basis ready for the sampling phase. Table 3.1 also shows the results for warm-starting all subs. The computing time remains approximately constant over the whole range of sample sizes. The results show that the jump in computing time for $n > n_{\text{prep}}$ is completely eliminated. No significant time differences can be observed. Thus the model for determining the parallel time t_p is valid for all numbers of preparatory problems n_{prep}.

The analysis so far has concerned previous implementations where the assignment of subproblems to subprocessors was hardwired. In our current implementation, subproblems are sent to the next idle node. This implementation allows for any number of subproblems n_{prep} and n per iteration and divides up the number of subproblems efficiently among the processors available. If necessary, the warm start procedure is used. In the following we are interested in the efficiency of the method with respect to both the sample size and the number of processors.

For determining efficiency we use the formulas developed in the previous section. Varying the sample size over a sufficiently large range, we estimate the parameters for determining the computing time. Table 3.2 gives results for sample sizes from 100 to 600 using 64 processors; it shows the parallel computation time versus the sample size for both the actual time measurements and the estimates from the formulas. One can see that the algebraic formulas give an excellent estimate of the actual parallel computing time. We estimated $t_{\text{MA}} + t_{\text{SUB}}^0$ to be 0.0962 and t_{SUB} to be 0.0149. Using these parameters we compute the corresponding serial time t_s, the speedup S, and the efficiency E, which are also reported in Table 3.2. The efficiency is low for small sample sizes, but it rapidly improves with increased sample size. In the case of sample size 600, we obtained a speedup of about 37.5, which means that with 64 processors we reduce the computation time by a factor of 37.5. The total parallel time was 17.3 minutes, whereas in a serial implementation the time to solve the problem would be 652 minutes. Figure 3.3 shows the relation of efficiency to sample size when 64 processors are used.

Table 3.2: Speedup and Efficiency

n	Iter.	t_p	t_p	t_s	S	E
		Actual	Est. by formula		Speedup	Efficiency
100	63	0.132	0.135	2.024	14.99	23.456
200	72	0.159	0.159	3.519	22.13	34.674
300	76	0.182	0.182	5.014	27.55	42.973
400	84	0.213	0.206	6.508	31.59	49.360
500	69	0.229	0.230	8.003	34.80	54.428
600	69	0.250	0.253	9.497	37.54	58.547

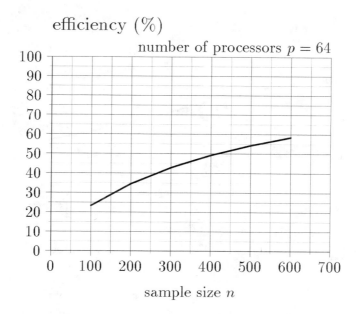

Figure 3.3: Efficiency versus sample size

Using estimates based on the formulas for the parallel time, we compute efficiency as a function of the number of processors used (for a fixed sample size). Figure 3.4 gives a graphical representation. For a small number of processors, the effect of having only $p-1$ processors operating in parallel when using p processors dominates the result. For example, when using two processors, we switch between the master processor and only one subprocessor. There is no parallel overlapping in the computation. In this case we perform a serial computation distributed to two processors. The efficiency is hence 50%. The efficiency increases until the above-mentioned effect is no longer dominating. For example, for sample size 600 and 12 processors, the efficiency is about 82%. The efficiency decreases with increasing numbers of processors beyond 12. Using 64 processors, we obtain an efficiency of 58.54% when the sample size is 600.

As an indication of the parallel implementation's potential, we estimated the efficiency using our formulas for much larger sample sizes. For example, we

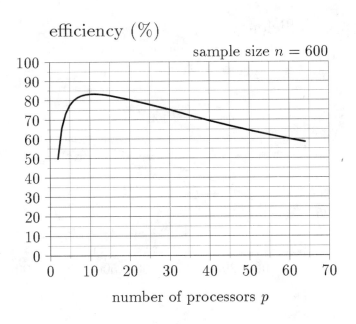

Figure 3.4: Efficiency versus number of processors

obtained an efficiency of 91% for a sample size of 5000, and for a sample size of 10,000 we obtained an efficiency of 95%. The efficiency obtained is high even though our implementation does not use asynchronous tune-up techniques, such as passing nonoptimal first-stage variables to the second stage in order to prevent the subprocessors from waiting until the master processor has finished solving the master problem, and other schemes. As one can easily see from the formulas, multiplying $t_{MA}+t_{SUB}^0$ and t_{SUB} by a constant factor does not affect the resulting efficiency. That means the efficiency is unaffected by proportional changes in the sizes of the master problem and subproblems. On the other hand, more expensive subproblems increase the efficiency. Thus, large sample sizes and large subproblems lead to increasing efficiencies and make the parallel implementation well suited to solving very large stochastic programs.

For the runs documented in Table 3.2, Table 3.3 reports on the optimum objective function value and the 95% confidence interval. The lower bound distributions have less variance than the upper bound distributions; hence the confidence interval is asymmetric. Using a sample size of 100 (out of about 1 million universe scenarios), we obtain an optimal solution of 188,348.7 with a 95% confidence interval of 0.08% on the lower side and 0.18% on the upper side. Even with only small sample sizes we obtain highly accurate results. The parallel time to run the problem was 8.3 minutes on the Hypercube.

The optimal objective function value remains stable when the sample size is increased. That again shows that we obtained good estimates. The confidence

Table 3.3: Optimal Solution

n	Iter.	Obj.	95% confidence interval			%	CPU
			lower	upper	total	of obj.	(min)
					lower +upper		
100	63	188348.7	153.0	344.4	497.4	0.26	8.3
200	72	188390.9	144.8	161.8	306.6	0.16	11.4
300	76	188344.9	100.5	180.2	280.7	0.15	13.8
400	84	188328.4	79.9	153.7	233.5	0.12	17.9
500	69	188304.0	78.0	131.1	209.0	0.11	15.8
600	69	188351.8	75.1	121.2	196.3	0.10	17.3

interval decreases with increasing sample size, and the rate of $n^{-0.5}$ is verified by the computational results.

Using a sample size of 600, we obtain an optimal objective function value of 188,351.8 with a 95% confidence interval of 0.04% on the lower side and 0.06% on the upper side. Thus the optimal solution with 95% confidence satisfies $188{,}276.7 \leq z^* \leq 188{,}473.0$. All solutions reported in Table 3.3 fall within this range. The computation time on the Hypercube was 17.3 minutes. It is interesting to note that during the process of solving the problem about 43,400 subproblems, with 289 rows and 302 columns each, and 69 master problems were solved.

3.5 SUMMARY—PARALLEL PROCESSING

We have demonstrated how Benders decomposition and importance sampling for solving two-stage stochastic linear programs can be implemented on a (Hypercube) parallel multicomputer and have presented numerical results for large-scale test problems. The efficiency of the implementation depends on the number of subproblems solved per iteration and the number of processors used. Large sample sizes and large subproblems lead to increasing efficiencies and make a parallel implementation well suited to solving very large stochastic programs. Current research involves the implementation of our methodology on clusters of powerful workstations. Future research should include experiments with asynchronous techniques for passing nonoptimal solutions in order to further optimize the efficiency of using parallel processors.

4

Techniques
for Solving
Multi-Stage
Problems

4.1   MULTI-STAGE
STOCHASTIC LINEAR
PROGRAMS

The use of stochastic programming techniques for solving large dynamic systems under uncertainty has been hampered until recently by the sheer size of the problems constructed when they are restated as deterministic linear programs. To solve them it has been necessary to keep the number of scenarios representing uncertainties fairly small. Only a few attempts have been made to solve practical multi-stage decision models whose future events are spread over several periods. In this section we extend the methodology developed for solving two-stage problems to multi-stage problems.

    Multi-stage planning problems can often be formulated as linear programs with a dynamic matrix structure which, in the deterministic case, appears in a staircase pattern of blocks formed by nonzero submatrices. These blocks correspond to and are different for different time periods. System (4.1) below shows the structure of a dynamic linear program of T periods.

$$\min\ z\ =$$

$$c_1 x_1\ +\ c_2 x_2\ +\cdots+\ c_{T-1} x_{T-1}\ +\ c_T x_T$$

s/t

$$
\begin{aligned}
A_1 x_1 &\qquad\qquad\qquad\qquad\qquad\qquad\quad = b_1\\
-B_1 x_1\ +\ A_2 x_2 &\qquad\qquad\qquad\qquad\qquad\quad = b_2\\
&\ddots\qquad\qquad\qquad\qquad\qquad\ \ \vdots\\
&\quad -B_{T-1} x_{T-1}\ +\ A_T x_T\ =\ b_T\\
x_1\ ,\qquad x_2\ ,\ldots, &\qquad\quad x_{T-1}\ ,\qquad x_T\ \geq\ 0.
\end{aligned}
$$

(4.1)

In the stochastic case, the blocks of coefficients and right-hand sides in different time periods are functions of several parameters whose values vary stochastically according to dependent and independent distributions which we assume to be known. The resulting problem is a multi-stage stochastic linear program. Even for a problem with a small number of stochastic parameters per stage, the size of the multi-stage problem, when expressed in equivalent deterministic form, can get so large as to appear intractable.

The multi-stage stochastic extension of a deterministic dynamic linear program can be formulated as follows:

$$\min\ z\ =$$

$$c_1 x_1\ +\ E(c_2 x_2^{\omega_2}\ +\cdots+\ E(c_{T-1} x_{T-1}^{\omega_{T-1},\ldots,\omega_2}\ +\ E(c_T x_T^{\omega_T,\ldots,\omega_2})))$$

s/t

$$
\begin{aligned}
A_1 x_1 &\qquad\qquad\qquad\qquad\qquad\qquad\qquad\qquad = b_1\\
-B_1^{\omega_2} x_1\ +\ A_2 x_2^{\omega_2} &\qquad\qquad\qquad\qquad\qquad\qquad\quad = b_2^{\omega_2}\\
&\ddots\qquad\qquad\qquad\qquad\qquad\qquad\qquad \vdots\\
&-B_{T-1}^{\omega_T} x_{T-1}^{\omega_{T-1},\ldots,\omega_2}\ +\qquad A_T x_T^{\omega_T,\ldots,\omega_2}\ =\ b_T^{\omega_T}\\
x_1\ ,\qquad x_2^{\omega_2}\ ,\ldots, &\qquad x_{T-1}^{\omega_{T-1},\ldots,\omega_2}\ ,\qquad x_T^{\omega_T,\ldots,\omega_2}\ \geq\ 0
\end{aligned}
$$

(4.2)

$$\omega_t \in \Omega_t,\ t = 2,\ldots,T.$$

The first-stage parameters c_1, A_1, and b_1 are known to the planner with certainty, but the parameters of stages $2, \ldots, T$ are assumed to be known only by their distributions. We consider uncertainty in the coefficients of the transition matrices $B_t^{\omega_t}$, $t = 1, \ldots, T-1$ and of the right-hand sides $b_t^{\omega_t}$, $t = 1, \ldots, T$. We assume the coefficients of the technology matrices A_t, $t = 1, \ldots, T$, and the objective function coefficients c_t, $t = 2, \ldots, T$, to be known with certainty. The assumption of deterministic technology matrices and objective function coefficients eases the presentation but is not crucial to the solution method we have developed. The goal of the planner is to minimize the expected value of present and future costs.

In a general model, one has to allow the stochastic parameters of the transition matrices $B_t^{\omega_t}$ and the right-hand sides $b_t^{\omega_t}$ to be dependent both within a certain stage and between stages. In the latter case the distributions of the stochastic parameters in period t depend on the outcomes of the stochastic parameters in period $t - 1$.

The underlying "wait and see" decision-making process is as follows. The decision maker makes a first-stage decision $\hat{x}_1$ before observing any outcome of the random parameters. Then he waits until an outcome of the second-stage random parameters is realized. The second-stage decision is then made based on the knowledge of the realization ω_2 but without observing the outcome of any random parameters of stages $3, \ldots, T$, and so forth. As the state (the actual outcome) is carried forward to the following period, the decision tree grows exponentially with the number of stages. The corresponding decision tree is given in Figure 4.1 for a four-stage problem with only three outcomes per stage.

We consider discrete distributions of random parameters with a finite number of outcomes, e.g., $\omega_t \in \Omega_t$, $\Omega_t = \{1, \ldots, K_t\}$, $t = 1, \ldots, T$, with corresponding probabilities $p_t^{\omega_t}$. With K_t being the number of scenarios in period t, the total number of scenarios for all T stages is $\prod_{t=1}^{T} K_t$. The number K_t is expected to be large because it is derived from crossing the sets of possible outcomes of the different random parameters within a period. For example, if the dimension of the random vector in period t is h_t and Ω_t^j contains k_t^j elements, then $K_t = \prod_{j=1}^{h_t} k_t^j$.

For example, in the hydro power control problem of Section 4.8.1, consider the case of prediction error in 10 inflows, modeled as independent random parameters with three outcomes each: the number of scenarios per period is $3^{10} \approx 60,000$. If we consider only 3 periods of "wait and see" planning ahead and neglect uncer-

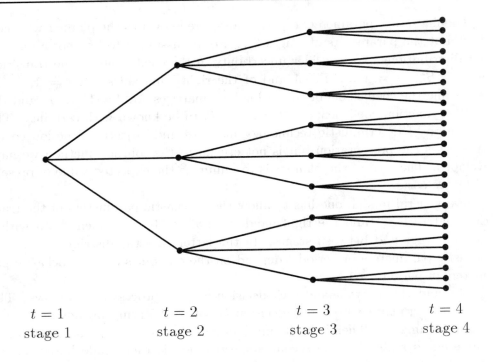

$t = 1$
stage 1

$t = 2$
stage 2

$t = 3$
stage 3

$t = 4$
stage 4

Figure 4.1: Decision tree of a general class of "wait and see" multi-stage problems

tainties of periods further ahead, then the total number of scenarios that have to be examined is on the order of 10^9.

As another example, in the asset allocation problem of Section 4.8.2, consider the case of 20 independent factors (used to describe the returns of, say, 500 assets) modeled as random parameters with five outcomes each: the number of scenarios per period is $5^{20} \approx 10^{14}$. If there are only three periods, the total number of scenarios grows to 10^{28}. The dimensions of an equivalent linear program of an asset allocation problem with a universe of about 500 assets is approximately $5 \cdot 10^{30}$ rows and $1.5 \cdot 10^{31}$ columns. It is, of course, impossible to write down this linear program explicitly.

4.2 DEPENDENCY AMONG STOCHASTIC PARAMETERS

We model dependency between stochastic parameters both within a certain stage and between stages. We refer to the former kind of dependency as **intra-stage** dependency and to the latter as **inter-stage** dependency.

We define the vector W_t to be the vector of all dependent random parameters of stage t. W_t contains both the random parameters in the transition matrix B_{t-1} and the random parameters of the right-hand side b_t. We will use the notation

$$W_t = \mathrm{vec}(B_{t-1}, b_t) \tag{4.3}$$

to express that all random parameters in the transition matrix B_{t-1} and all random parameters in the right-hand side b_t are organized in the random vector W_t. Each component of W_t refers uniquely to an element of the matrix B_{t-1} or to an element of the vector b_t. We define inverse operations of $W_t = \mathrm{vec}(B_{t-1}, b_t)$ as

$$B_{t-1}(W_t) = \mathrm{vec}_B^{-1}(W_t) \tag{4.4}$$

and

$$b_t(W_t) = \mathrm{vec}_b^{-1}(W_t) \tag{4.5}$$

to express that, given the random vector W_t, the corresponding random elements in B_{t-1} and b_t are uniquely determined. With $m_{t+1} \times n_t$ as the dimensions of the transition matrices B_t and m_t as the dimension of the right-hand side b_t, the dimension of the dependent random vector W_t is at most $m_{t+1} \times n_t + m_{t+1}$. We denote the dimension of W_t with $\tilde{h}_t$; thus $W_t = (W_{1,t}, \ldots, W_{\tilde{h}_t,t})$. With respect to the solution procedure discussed later, we formulate a special kind of dependency model that is based on linear intra-stage and inter-stage dependency.

4.2.1 Intra-Stage Dependency

We describe correlation of random parameters within a stage by a linear relation; i.e., outcomes of the $\tilde{h}_t$ uncertain dependent parameters $(W_{1,t}, \ldots, W_{\tilde{h}_t,t})$ can be obtained by multiplying outcomes of a vector of independent random parameters $V_t = (V_{1,t}, \ldots, V_{h_t,t})$ by a matrix $\tilde{\beta}_t(\tilde{h}_t \times h_t)$:

$$W_t = \tilde{\beta}_t V_t. \qquad (4.6)$$

We denote an outcome of the independent random vector V_t by v_t or $v_t^{\omega_t}$, where $\omega_t \in \Omega_t$. We denote an outcome of an individual random parameter in period t, say $V_t^{j_t}$, by $v_t^{j_t}$, also denoted as v^{j_t,ω_t^j}, where $\omega_t^{j_t} \in \Omega_t^{j_t}$, with corresponding probability $p(v_t^{j_t}) = \text{prob}\,(V_t^{j_t} = v_t^{j_t})$. The set of all possible outcomes in stage t is constructed by crossing the sets of possible outcomes of the components of V_t, $\Omega_t^{j_t}$, $j_t = 1, \ldots, h_t$, as $\Omega_t = \Omega_t^1 \times \Omega_t^2 \times \cdots \times \Omega_t^{h_t}$. Because of the independence of the individual random parameters of V_t, $p(v_t) = p(v_t^1)p(v_t^2)\cdots p(v_t^{h_t})$. The corresponding outcomes of the dependent random vectors W_t are denoted by w_t or $w_t^{\omega_t}$.

The linear intra-stage dependency model generates dependent scenarios based on random variations of independent random parameters. This property will become particularly important when implementing sampling for generating scenarios. In many applications the number of independent random parameters per stage, h_t, is expected to be much smaller than the number of dependent random parameters, $\tilde{h}_t$.

4.2.2   Inter-Stage Dependency

For expository purposes we consider an equal number of independent random parameters in each of the different stages and denote this number as h. We consider inter-stage dependency as a Markovian process (linear dependency):

$$w_t^{\omega_t,\ldots,\omega_2} = \beta_0 v_t^{\omega_t} + \beta_1 w_{t-1}^{\omega_{t-1}} + \cdots + \beta_{t-2} w_2^{\omega_2}. \qquad (4.7)$$

The matrices $\beta_0, \beta_1, \ldots, \beta_{T-2}$ are matrices of dimension $(h \times h)$. If they are diagonal matrices, the value of an outcome of the random parameter W_t^i is a weighted sum of some independent random variation in period t, denoted by v_t^i, and the values of the observed outcomes in the previous periods $t-1, t-2, \ldots, 2$. If the matrices $\beta_0, \beta_1, \ldots, \beta_{T-2}$ are nondiagonal, the outcome of each element in

W_t depends on the outcomes of all elements in W_{t-1}, W_{t-2}, back to W_2 and the independent random variations of V_t. Given observed outcomes of W_{t-1}, W_{t-2}, back to W_2, one can easily generate a scenario of W_t by independent random variations of V_t. One can see that a scenario in period t is determined based on the observed historical outcomes of the dependent random parameters and by the variations of the independent random parameters in period t. We refer to our linear dependency model as an **auto-correlative model**. Auto-correlative models constitute a wide class of dependency models that is well studied in the literature. Many different applications in practice can be formulated using the auto-correlative approach. The parameters of the auto-correlative model can be estimated based on historical observations of the outcomes of the stochastic parameters. We can now state the auto-correlative model in the notation of the formulation of the multi-stage stochastic linear program:

$$B_{t-1}(w_t^{\omega_t,\dots,\omega_2}) = \mathrm{vec}_B^{-1}(\beta_0 v_t^{\omega_t}) + \sum_{\tau=1}^{t-2} \mathrm{vec}_B^{-1}(\beta_\tau \mathrm{vec}(B_{t-\tau-1}^{\omega_{t-\tau}}, b_{t-\tau}^{\omega_{t-\tau}})), \qquad (4.8)$$

$$b_t(w_t^{\omega_t,\dots,\omega_2}) = \mathrm{vec}_b^{-1}(\beta_0 v_t^{\omega_t}) + \sum_{\tau=1}^{t-2} \mathrm{vec}_b^{-1}(\beta_\tau \mathrm{vec}(B_{t-\tau-1}^{\omega_{t-\tau}}, b_{t-\tau}^{\omega_{t-\tau}})). \qquad (4.9)$$

The general auto-correlative model of equations (4.8) and (4.9) above covers a wide range of models reflecting additive dependency. Special cases, where uncertainty appears either in the right-hand sides b_t or in the transition matrices B_{t-1} but not in both at the same time, are common practical examples.

4.3   DUAL DECOMPOSITION

A description of how the Benders (1962) [6] decomposition algorithm can be applied to solve two-stage stochastic linear programs can be found in Van Slyke and Wets (1969) [122]. Birge (1985) [10] extended the two-stage concept to solving multi-stage stochastic linear programs by using a nested Benders decomposition scheme.

Using dual decomposition, we decompose the problem into subproblems for different stages t. In the most general case, in which there is a dependency of stochastic parameters between stages, the number of subproblems is equal to the number of scenarios in each stage t. To distinguish one subproblem from another, each is indexed with $\omega_t, \ldots, \omega_2$, where ω_t is the random event in stage t and $\omega_{t-1}, \ldots, \omega_2$ is the path of previous events that gave rise to the particular subproblems in stage t.

For expository purposes, we assume initially that the random events occurring in one stage are independent of those that happened in the previous stage. In the independent case, scenarios $\omega_{t+1} \in \Omega_{t+1}$ in period $t+1$ are identical for each scenario $\omega_t \in \Omega_t$ in period t. The history is only carried forward through optimal decisions $\hat{x}_{t-1}^{\omega_{t-1}, \ldots, \omega_2}$ from previous periods.

The idea of using dual decomposition is to express in each stage t, and scenario ω_t, the expected future costs (the impact of stages $t+1, \ldots, T$) by a scalar θ_t. The value of θ_t is constrained by a set of cuts, which are necessary conditions for feasibility and optimality expressed solely in terms of the stage t decision variables x_t and θ_t. Cuts are initially absent and then sequentially added to the stage t problems. Each scenario subproblem collects the information about expected future costs by means of the cuts.

The relationship between the different stages and scenarios in the decomposed multi-stage problem is summarized as follows.

The stage 1 problem:

$$
\begin{aligned}
\min \; z_1 \; = \quad & c_1 x_1 \; + \quad \theta_1 \\
\text{s/t} \quad & \\
\pi_1 : \quad & A_1 x_1 \qquad\qquad\quad = \; b_1 \\
\rho_1^{l_1} : \quad & -G_1^{l_1} x_1 \; + \; \alpha_1^{l_1} \theta_1 \; \geq \; g_1^{l_1}, \quad l_1 = 1, \ldots L_1 \\
& x_1 \qquad\qquad\qquad\quad \geq \; 0.
\end{aligned}
\tag{4.10}
$$

The stage t, $t = 2, \ldots, T - 1$, problems:

$$
\begin{aligned}
\min \; z_t^{\omega_t} \; = \quad & c_t x_t^{\omega_t} \; + \quad \theta_t^{\omega_t} \\
\text{s/t} \quad & \\
\pi_t^{\omega_t} : \quad & A_t x_t^{\omega_t} \qquad\qquad\quad = \; b_t^{\omega_t} + B_{t-1}^{\omega_t} \hat{x}_{t-1} \\
\rho_t^{l_t, \omega_t} : \quad & -G_t^{l_t} x_t^{\omega_t} \; + \; \alpha_t^{l_t} \theta_t^{\omega_t} \; \geq \; g_t^{l_t}, \quad l_t = 1, \ldots, L_t \\
& x_t^{\omega_t} \qquad\qquad\qquad\quad \geq \; 0.
\end{aligned}
\tag{4.11}
$$

The stage T problem:

$$
\begin{aligned}
\min \; z_T^{\omega_T} &= c_T x_T^{\omega_T} \\
\text{s/t} & \\
\pi_T^{\omega_T} : \qquad A_T x_T^{\omega_T} &= b_T^{\omega_T} + B_{T-1}^{\omega_T} \hat{x}_{T-1} \\
x_T^{\omega_T} &\geq 0.
\end{aligned}
\tag{4.12}
$$

$\min z_1$ represents the optimal objective function value in the first stage, and x_1 and θ_1 represent the optimal solution. The vector π_1 denotes the optimal dual prices associated with the original first-stage constraints, and the scalars $\rho_1^{l_1}$ are the optimal dual prices associated with the cuts that have been added so far in decomposition iterations $l_1 = 1, \ldots, L_1$. In later stages, the optimal objective function values $\min z_t^{\omega_t} = \min z_t^{\omega_t}(\hat{x}_{t-1})$, the optimal dual prices $\pi_t^{\omega_t} = \pi_t^{\omega_t}(\hat{x}_{t-1})$ associated with the original stage t constraints, and the optimal dual prices $\rho_t^{l_t,\omega_t} = \rho_t^{l_t,\omega_t}(\hat{x}_{t-1})$ associated with the cuts are all dependent upon $\hat{x}_{t-1}$, the optimal solution passed as input from the previous stage $t-1$. According to the scenario evolution in the previous stages, an optimal solution $\hat{x}_{t-1}$ is actually indexed by the scenario outcomes of all previous stages and is therefore denoted as $\hat{x}_{t-1}^{\omega_{t-1},\ldots,\omega_2}$. For the sake of exposition, we suppress the scenario history and present the optimal solution of subproblems in stage t, scenario ω_t as a function of the input $\hat{x}_{t-1}$.

We compute the expected future costs as

$$
z_{t+1}(\hat{x}_t^{l_t}) = E_{\omega_{t+1}} z_{t+1}^{\omega_{t+1}}(\hat{x}_t^{l_t}),
\tag{4.13}
$$

the expected right-hand sides of the cuts as

$$
g_t^{l_t} = E_{\omega_{t+1}} \left(\pi_{t+1}^{\omega_{t+1}}(\hat{x}_t^{l_t}) b_{t+1}^{\omega_{t+1}} + \sum_{l_{t+1}=1}^{L_{t+1}} \rho_{t+1}^{l_{t+1},\omega_{t+1}}(\hat{x}_t^{l_t}) g_{t+1}^{l_{t+1}} \right),
\tag{4.14}
$$

and the expected coefficients of the cuts as

$$
G_t^{l_t} = E_{\omega_{t+1}} \pi_{t+1}^{\omega_{t+1}}(\hat{x}_t^{l_t}) B_t^{\omega_{t+1}},
\tag{4.15}
$$

where

$$\rho_T^{\omega_T} = 0, \ G_T^{\omega_T} = 0, \ and \ g_T^{\omega_T} = 0. \tag{4.16}$$

A subproblem in stage t and in scenario ω_t interacts with its predecessors and descendants by passing cuts backward and optimal solutions forward. Dual decomposition splits the multi-stage problem into a series of two-stage relations that are connected overall by a nesting scheme. We call the stage t, scenario ω_t problem the current master problem. It receives from its ancestor in period $t - 1$ a solution $\hat{x}_{t-1}$. The current scenario is determined by the outcome ω_t of the random parameters in stage t, which are reflected in the right-hand side $b_t^{\omega_t} + B_{t-1}^{\omega_t}\hat{x}_{t-1}$. As stated above, $\hat{x}_{t-1}$ imbeds a history. This history has to be considered when several stages are nested. Given and subject to $\hat{x}_{t-1}$, we solve the stage t problem in scenario ω_t and pass the obtained solution $\hat{x}_t^{\omega_t}$ to the descendant problems. By solving all problems $\omega_{t+1} \in \Omega_{t+1}$ (referred to as the universe case), we can compute exactly the expected value of the descendant-stage costs z_{t+1} and the coefficients $G_t^{l_t}$ and right-hand side $g_t^{l_t}$ of a cut. The cut is added to the current master problem (stage t, scenario ω_t), and by solving this problem again, we obtain another trial solution.

The optimal trial solution of the current master problem in stage t, scenario ω_t gives a lower bound, and the expected cost of this trial solution gives an upper bound for the expected costs of all scenarios descendant from stage t, scenario ω_t. If the lower and upper bounds are sufficiently close, the current master problem is deemed to represent adequately the future expected costs and it contains (by means of a sufficient number of cuts) all the information needed from future scenarios. In this case we say **the current master is balanced** with its descendant problems.

Note that the current master problem represents only the expected future costs subject to the trial solution $\hat{x}_{t-1}$, which was passed from its ancestor, and subject to the current scenario ω_t. Furthermore, we have implicitly assumed that the descendant problems in stage $t + 1$ are also balanced with their descendant problems. Note, however, that the solution of the current stage t, scenario ω_t problem gives a lower bound for the expected costs of all descendant scenarios regardless of having collected a sufficient number of cuts. We shall exploit this fact.

4.4 PROPERTIES OF THE CUTS

First we discuss various properties of cuts in the case of independence of stochastic parameters between stages. Then we extend the discussion to the case of Markovian dependency and restate the derived properties for the dependent case.

4.4.1 The Case of Independence of Stochastic Parameters between Stages

The following properties of cuts are crucial for the solution procedure.

Cuts derived from any trial solution $\hat{x}_t^{\omega_t}$ are valid in all scenarios $\omega_t \in \Omega_t$.

The cut $\theta_t \geq E_{\omega_{t+1}} \pi_{t+1}^{\omega_{t+1}} B_t^{\omega_{t+1}} x_t + E_{\omega_{t+1}} (\pi_{t+1}^{\omega_{t+1}} b_{t+1}^{\omega_{t+1}} + \sum_{l_{t+1}=1}^{L_{t+1}} \rho_{t+1}^{l_{t+1},\omega_{t+1}} g_{t+1}^{l_{t+1}})$ is a constraint whose coefficients do not depend on x_t; hence, it is valid for all values of x_t. We exhibit this property via the example of cuts generated for the stage $T-1$ master problem and show that these cuts are valid for all values $x_{T-1}^{\omega_{T-1}}$. We employ dual (Benders) decomposition theory to obtain the relation between the stage $T-1$ master problem and the stage T subproblems. First we restate the stage $T-1$ master problem of the multi-stage stochastic linear program.

The stage $T-1$ relaxed master problem:

$$
\begin{aligned}
z_{T-1} \quad &= \quad \min \; c_{T-1} x_{T-1} \; + \; z_T(x_{T-1}) \\
&\text{s/t} \quad (A_{T-1} x_{T-1} \qquad\qquad = \quad b_{T-1}^{\omega_{T-1}} + B_{T-2}^{\omega_{T-1}} \hat{x}_{T-2}) \qquad (4.17) \\
&\qquad\quad x_{T-1} \qquad\qquad\qquad\qquad \geq \quad 0.
\end{aligned}
$$

System (4.17) represents the relaxed stage $T-1$ master problem, where no cuts have been added so far. As an initial step, we also relax the stage $T-1$ original constraints $(A_{T-1} x_{T-1} = b_{T-1}^{\omega_{T-1}} + B_{T-2}^{\omega_{T-1}} \hat{x}_{T-2})$, as indicated by setting them between parentheses. This is to make the stage $T-1$ problem independent of the different scenarios ω_{T-1} and independent of any history that is represented by $\hat{x}_{T-2}$ passed from the previous stage. Thus we can interpret the stage $T-1$ master problem as being completely cut off from the previous stages. The expected (future) stage T costs, $z_T(x_{T-1})$, are represented as a function of the stage $T-1$ decision variable x_{T-1}.

Instead of stating the primal stage T problem, we state its dual problem. The dual stage T problem in (4.18) includes all K_T scenarios $\omega_T \in \Omega_T$. Note that the stage $T-1$ decision, $\hat{x}_{T-1}$, appears as a given parameter of the stage T subproblem.

The stage T dual subproblem:

$z_T(\hat{x}_{T-1}) =$

$$\max \quad p_T^1 \pi_T^1 (b_T^1 + B_{T-1}^1 \hat{x}_{T-1}) \;\; + \cdots + \;\; p_T^{K_T} \pi_T^{K_T} (b_T^{K_T} + B_{T-1}^{K_T} \hat{x}_{T-1})$$

$$\text{s/t}$$

$$\pi_T^1 A_T \qquad\qquad\qquad\qquad\qquad\qquad\qquad \le \;\; c_T$$

$$\ddots$$

$$\pi_T^{K_T} A_T \qquad\qquad\qquad \le \;\; c_T.$$

$$(4.18)$$

We assume the stage T dual subproblem to be finite for all values of $\hat{x}_{T-1}$ passed from the previous period. We define

$$\pi_T^j := (\pi_T^{1,j}, \ldots, \pi_T^{K_T,j}), \quad j = 1, \ldots, q_T \qquad\qquad (4.19)$$

to be the vertices of the dual feasible region of the stage T subproblem (4.18). We rewrite the stage T subproblem (4.18) by expressing it in terms of the dual vertices. By doing so, we can also write the minimum expected costs $z_T(x_{T-1})$ as a function of x_{T-1}, rather than having the problem parameterized by $\hat{x}_{T-1}$:

$$z_T(x_{T-1}) = \max_{1 \le j \le q_T} \quad p_T^1 \pi_T^{1,j} (b_T^1 + B_{T-1}^1 x_{T-1}) + \cdots + p_T^{K_T} \pi_T^{K_T,j} (b_T^{K_T} + B_{T-1}^{K_T} x_{T-1}).$$

$$(4.20)$$

Using θ_{T-1} to represent the smallest upper bound for the stage T costs,

$$\min \theta_{T-1},$$
$$\theta_{T-1} \geq \max_{1 \leq j \leq q_T} \; p_T^1 \pi_T^{1,j}(b_T^1 + B_{T-1}^1 x_{T-1}) + \cdots + p_T^{K_T} \pi_T^{K_T,j}(b_T^{K_T} + B_{T-1}^{K_T} x_{T-1}),$$

$$(4.21)$$

we can state the stage $T-1$ master problem as a full master problem with all possible cuts added.

The stage $T-1$ full master problem:

$$\min z_{T-1} = c_{T-1} x_{T-1} + \theta_{T-1}$$
$$\text{s/t} \quad (A_{T-1} x_{T-1} = b_{T-1}^{\omega_{T-1}} + B_{T-2}^{\omega_{T-1}} \hat{x}_{T-2}) \qquad (4.22)$$
$$x_{T-1} \geq 0$$

$$\theta_{T-1} \geq p_T^1 \pi_T^{1,j}(b_T^1 + B_{T-1}^1 x_{T-1}) + \cdots + p_T^{K_T} \pi_T^{K_T,j}(b_T^{K_T} + B_{T-1}^{K_T} x_{T-1}) \quad j = 1, \ldots, q_T.$$

$$(4.23)$$

In the full master representation for stage $T-1$, each cut corresponds to a vertex of the dual feasible region of the stage T subproblem. The set of all cuts represents a complete outer linearization of the expected stage T costs as a function of x_{T-1}. Each cut is valid for each value of x_{T-1} because it is derived from a dual feasible price vector, one of the dual vertices π_T^j, $j = 1, \ldots, q_T$. Taking now the stage $T-1$ constraints $(A_{T-1} x_{T-1} = b_{T-1}^{\omega_{T-1}} + B_{T-2}^{\omega_{T-1}} \hat{x}_{T-2})$ into consideration, we see that the piecewise-linear representation of the expected stage T costs remains unaffected by the stage $T-1$ constraints. The stage $T-1$ constraints have different right-hand sides in each scenario ω_{T-1} and for each value of $\hat{x}_{T-2}$ passed from the previous period. Therefore they represent a different set of feasible solutions for x_{T-1} in each scenario ω_{T-1} and for each value of $\hat{x}_{T-2}$. The piecewise-linear representation of the stage T expected costs by means of the cuts is valid for any scenario ω_{T-1} and any value of $\hat{x}_{T-2}$ passed from the previous period. Of course, different cuts will be binding in different scenarios

ω_{T-1}. For example, a cut can be a support of the recourse function, the minimum expected costs $z_T(x_{T-1})$ as a function of the stage $T-1$ decision variables x_{T-1}, in a particular scenario and may be well below the recourse function in another scenario.

Having stated the full master problem in stage $T-1$ in equations (4.22 – 4.23), we now extend the argument for all stages. As before, we state the dual problems in stages t, where $t = 2, \ldots, T-1$. In order to ease the following presentation, we will use the matrix G_t to represent the matrix of coefficients of all cuts $G_t^{l_t}$, $l_t = 1, \ldots, L_t$, where each $G_t^{l_t}$ is a row of G_t, and the column vector g_t to represent the right-hand sides of the cuts $g_t^{l_t}$, $l_t = 1, \ldots, L_t$. Similarly we define as $\rho_t^{\omega_t}$ the vector of dual variables $\rho_t^{l_t, \omega_t}$, $l_t = 1, \ldots, L_t$. We assume that the subproblem in stage t under consideration contains a complete representation of the stage $t+1$ expected costs, represented by the set of cuts $-G_t^{l_t} x_t^{\omega_t} + \theta_t^{\omega_t} \geq g_t^{l_t}$, $l_t = 1, \ldots, L_t$, where $L_t = q_{t+1}$, and that these cuts have also been derived from subproblems with a full representation of their respective future expected costs.

The stage t, $t = 2, \ldots, T-1$ dual subproblem:

$z_t(\hat{x}_{t-1}) =$

$$\max p_t^1(\pi_t^1(b_t^1 + B_{t-1}^1 \hat{x}_{t-1}) + \rho_t^1 g_t) + \cdots + p_T^{K_t}(\pi_t^{K_t}(b_t^{K_t} + B_{t-1}^{K_t}\hat{x}_{t-1}) + \rho_t^{K_t} g_t)$$

s/t

$$
\begin{array}{rcl}
\pi_t^1 A_t \quad -\rho_t^1 G_t & \leq & c_t \\
\ddots & & \\
\pi_t^{K_t} A_t \quad -\rho_t^{K_t} G_t & \leq & c_t \qquad (4.24) \\
\rho_t^1 1 & = & 1 \\
\ddots & & \\
& & \\
\rho_t^{K_t} 1 & = & 1 \\
\rho_t^1 \quad , \ldots, \qquad \rho_t^{K_t} & \geq & 0.
\end{array}
$$

As before, the stage $t-1$ master problem in the full master representation can be stated as

The stage $t - 1$ full master problem:

$$\min \; z_{t-1} \;=\; c_{t-1}x_{t-1} \;+\; \theta_{t-1}$$
$$\text{s/t} \quad (A_{t-1}x_{t-1} \;=\; b_{t-1}^{\omega_{t-1}} + B_{t-2}^{\omega_{t-1}}\hat{x}_{t-2}) \qquad (4.25)$$
$$x_{t-1} \;\geq\; 0$$

$$\theta_{t-1} \geq$$
$$p_t^1(\pi_t^{1,j}(b_t^1 + B_{t-1}^1 x_{t-1}) + \rho_t^{1,j} g_t) + \cdots + p_t^{K_t}(\pi_t^{K_t,j}(b_t^{K_t} + B_{t-1}^{K_t} x_{t-1}) + \rho_t^{K_t,j} g_t),$$
$$j = 1, \ldots, q_t.$$

$$(4.26)$$

Equation (4.26) represents a complete outer linearization of the future costs as a function of x_{t-1}. It is valid for all possible values of x_{t-1}. The stage $t - 1$ constraints have different right-hand sides for different scenarios ω_{t-1} and for different values of $\hat{x}_{t-2}$ passed from the previous stage. They represent a different set of feasible solutions of x_{t-1} in each scenario ω_{t-1} and for each value of $\hat{x}_{t-2}$, but the outer linearization of the stage $t + 1$ costs is valid in any scenario ω_{t-1} and for each value of $\hat{x}_{t-2}$ passed from the previous period. Therefore all cuts $-G_{t-1}^{l_{t-1}} x_{t-1}^{\omega_{t-1}} + \theta_{t-1}^{\omega_{t-1}} \geq g_{t-1}^{l_{t-1}}$, $l_{t-1} = 1, \ldots, L_{t-1}$, are valid for all scenarios $\omega_{t-1} \in \Omega_{t-1}$.

A cut derived from a nonextreme dual feasible point is a valid cut.

In the full master representation of the stage $t - 1$ problem in (4.25) and (4.26), each cut has been derived from a vertex of the dual feasible region of the stage t subproblem. A cut derived from a dual extreme point is efficient in the sense that it supports the expected future cost function $z_t(x_{t-1})$ at that value of $\hat{x}_{t-1}$ for which the dual extreme point is optimal. Of course, not every dual extreme point need be optimal for some value of x_{t-1}. A cut derived from a strict interior point of the dual feasible region cannot be a support of $z_t(x_{t-1})$ for any value of x_{t-1}, because such a point is never dual optimal. Therefore, it is not efficient in the above sense.

For example, any point of the feasible region of the stage t dual subproblem of (4.24) can be represented by a convex combination of the extreme points, if the feasible region is finite, e.g.,

$$(\pi_t^0, \rho_t^0) = \sum_{l_t=1}^{q_t} \lambda^{l_t}(\pi_t^{l_t}, \rho_t^{l_t}), \quad \sum_{l_t=1}^{q_t} \lambda^{l_t} = 1, \quad \lambda^{l_t} \geq 0. \tag{4.27}$$

Similarly, if the feasible region is infinite, any point of it can be represented by a linear combination of extreme points and extreme rays. We will, however, not discuss this case any further and continue with the finite case.

For any value of x_{t-1} the outer linearization of the expected stage t costs, based on the dual interior point π_t^0, ρ_t^0,

$$z_t^0(x_{t-1}) = E_{\omega_t}(\pi_t^{\omega_t,0}(b_t^{\omega_t} + B_{t-1}^{\omega_t}x_{t-1}) + \rho_t^{\omega_t,0}g_t), \tag{4.28}$$

is less than the outer linearization corresponding to the extreme point $\pi_t^{\max}, \rho_t^{\max}$ that maximizes the dual problem for a particular $\hat{x}_{t-1}$:

$$z_t^0(x_{t-1}) \leq E_{\omega_t}(\pi_t^{\omega_t,\max}(b_t^{\omega_t} + B_{t-1}^{\omega_t}x_{t-1}) + \rho_t^{\omega_t,\max}g_t), \tag{4.29}$$

where

$$E_{\omega_t}(\pi_t^{\omega_t,\max}(b_t^{\omega_t} + B_{t-1}^{\omega_t}x_{t-1}) + \rho_t^{\omega_t,\max}g_t) = \max_j E_{\omega_t}(\pi_t^{\omega_t,j}(b_t^{\omega_t} + B_{t-1}^{\omega_t}x_{t-1}) + \rho_t^{\omega_t,j}g_t). \tag{4.30}$$

It follows that a cut obtained from a dual feasible point that is not an extreme point of the dual feasible region in (4.24) is weak, but valid for any value of x_{t-1} of the stage $t-1$ master problem.

An extreme point of the dual feasible region of a subproblem that is not balanced with its descendant stage is also an extreme point of its dual feasible region when it is balanced with its descendant stage.

We analyze now the case where the stage t subproblem is not fully balanced with the stage $t + 1$ subproblem, that is, when not all cuts have been generated and added to the stage t subproblem to fully represent the expected stage $t + 1$ costs, $z_{t+1}(x_t)$, given $\hat{x}_{t-1}$ passed to the stage t problem and for all scenarios $\omega_t \in \Omega_t$ in stage t. To determine whether the cuts added so far to the stage t subproblem fully represent the stage $t + 1$ expected costs, we define $\theta^{\min}(x_t)$ as the smallest upper bound of the value of the stage $t + 1$ costs by means of the stage t cuts in the stage t subproblem:

$$\theta^{\min}(x_t) = \min \theta_t, \ \theta_t \geq G_t^{l_t} x_t + g_t^{l_t}, \ l_t = 1, \ldots, L_t. \tag{4.31}$$

If

$$\theta^{\min}(x_t) = z_{t+1}(x_t), \tag{4.32}$$

then the cuts added so far in stage t fully represent the stage $t + 1$ expected costs, and the stage t subproblem is balanced with the stage $t + 1$ subproblem. Otherwise, additional cuts have to be generated for stage t in order to obtain a complete representation of the stage $t + 1$ expected costs.

The case that the stage t subproblem is not balanced with the stage $t + 1$ subproblem can be interpreted in these terms: the dimension of the vector of dual prices ρ_t, corresponding to the number of cuts added so far to the stage t subproblem, is smaller than its dimension would be if all cuts were added to fully represent the expected future costs $z_{t+1}(x_t)$.

For example, the stage t full master problem has q_{t+1} extreme cuts but only L_t unique cuts have been added so far. The stage t subproblem can then be seen as having $q_{t+1} - L_t$ relaxed constraints, whose coefficients and right-hand sides are zero. The dual variables associated with these relaxed constraints are defined to have the value zero. In the dual representation of the stage t subproblem, these relaxed rows appear as zero-columns. An extreme point π_t^j, ρ_t^j of the dual feasible region corresponding to L_t cuts and with $q_{t+1} - L_t$ zero-columns is also an extreme point of the dual feasible region corresponding to q_{t+1} cuts and for which $\rho_t^j \geq 0$, for $j = L_t + 1, \ldots, q_{t+1}$. This is true because if $(\pi_t, \rho_t^1, \ldots, \rho_t^{L_t})$ is an extreme point of the relaxed dual, then the constructed dual solution $(\pi_t, \rho_t^1, \ldots, \rho_t^{L_t}, \rho_t^{L_t+1} = 0, \ldots, \rho_t^{q_t+1} = 0)$ could not be formed from a linear combination of other extreme points of the full system by virtue of the non-negativity of ρ_t. A cut obtained from a dual extreme point is efficient in the sense that it has the potential to be a support of the recourse function for some value of x_t. Therefore a cut obtained from stage $t+1$ subproblems that are not balanced with descendant subproblems in stage $t+2$ is an efficient cut. It is suboptimal for the particular value of $\hat{x}_t$ at which it was derived, but it may be optimal for another value of x_t. Note that this result has an important implication: no matter at which stage cuts are computed, regardless of whether the corresponding subproblems are balanced or unbalanced with their descendant stages, the obtained cuts are always efficient cuts. They are potentially supports of the recourse function at some value of x_t. Note that cuts computed in the first stage are efficient cuts, even if no cuts have been added in stages $2, \ldots, T - 1$, that is, if no information about the future is collected.

Cuts obtained from expected value subproblems are valid cuts.

By replacing all stochastic parameters $B_{t-1}^{\omega_t}$, $b_t^{\omega_t}$, $t = 2, \ldots, T$, with their expected values $B_{t-1}^E = E_{\omega_t}(B_{t-1}^{\omega_t})$, $b_t^E = E_{\omega_t}(b_t^{\omega_t})$, we obtain from the multi-stage stochastic linear program of (4.2) a deterministic multi-stage linear program (4.33), which we refer to as its expected-value problem.

The expected-value problem:

$$
\begin{aligned}
\min \ z = \quad & \\
c_1 x_1 \ + \ & c_2 x_2 \ + \cdots + \quad c_{T-1} x_{T-1} \ + \ c_T x_T \\
\text{s/t} \quad & \\
A_1 x_1 \qquad\qquad\qquad\qquad\qquad\qquad & = \ b_1 \\
-B_1^E x_1 \ + \ & A_2 x_2 \qquad\qquad\qquad\qquad = \ b_2^E \\
& \ddots \qquad\qquad\qquad\qquad\qquad \vdots \\
& \quad -B_{T-1}^E x_{T-1} \ + \ A_T x_T \ = \ b_T^E \\
x_1 \ , \quad & x_2 \ , \ldots, \qquad\quad x_{T-1} \ , \quad x_T \ \geq \ 0.
\end{aligned}
\tag{4.33}
$$

In order to demonstrate the properties of cuts obtained from expected-value subproblems, we apply dual (Benders) decomposition to (4.33). Corresponding to the stage t subproblem (4.11) of the multi-stage stochastic linear problem is the stage t subproblem (4.34) of the decomposed multi-stage expected-value problem.

The stage t expected-value subproblem:

$$
\begin{aligned}
\min\ z_t^E(\hat{x}_{t-1}) \ &=\ \min\ c_t x_t\ +\ \theta_t \\
&\text{s/t} \\
\pi_t^E:\qquad\quad A_t x_t\qquad\ &=\ b_t^E + B_{t-1}^E \hat{x}_{t-1} \qquad\qquad (4.34) \\
\rho_t^E:\quad\ -G_t^E x_t\ +\ \theta_t\ &\geq\ g_t^E \\
x_t\qquad\qquad &\geq\ 0.
\end{aligned}
$$

In the stochastic case we have defined the expected stage t costs as a function of x_{t-1} as

$$
z_t(x_{t-1}) = E_{\omega_t}(z_t^{\omega_t}(x_{t-1})), \qquad\qquad (4.35)
$$

where $z_t^{\omega_t}(x_{t-1})$ is a convex function. From this it follows (based on Jensen's inequality for convex functions) that

$$
z_t(x_{t-1}) \geq z_t^E(x_{t-1}). \qquad\qquad (4.36)
$$

Equation (4.36) is true for all values of x_{t-1}.

By the usual argument for Benders decomoposition, a cut derived from the expected-value subproblem is an outer linearization of the stage t expected-value subproblem costs $z_t^E(x_{t-1})$. To show this, we assume that problem (4.34) is feasible for all values of $\hat{x}_{t-1}$ and define $\pi_t^{E,j}, \rho_t^{E,j}, j = 1, \dots, q_t^E$ to be the extreme points of its dual feasible region, where q_t^E denotes the number of dual vertices. The outer linearization then is represented as

$$
L_t^E(x_{t-1}) = \max_j \pi_t^{E,j}(b_t^E + B_{t-1}^E x_{t-1}) + \rho_t^{E,j} g_t, \quad j = 1, \dots, q_t^E. \qquad (4.37)
$$

With

$$z_t^E(x_{t-1}) \geq L_t^E(x_{t-1}) \tag{4.38}$$

we can state

$$z_t(x_{t-1}) \geq L_t^E(x_{t-1}). \tag{4.39}$$

This is a valid statement for all values of x_{t-1} and is unaffected by any additional constraints on x_{t-1} in the stage $t-1$ subproblem. It follows from (4.36), then, that cuts obtained from the stage t expected-value subproblem are valid but weak cuts for any stage $t-1$, ω_{t-1} subproblem of the decomposed stochastic linear problem. Indeed, they are used to guide our solution algorithm at the beginning. For example, we first solve the expected-value problem (4.33) using nested dual decomposition and collect cuts at each stage t, $t = 1, \ldots, T-1$. Then we solve the stochastic problem, where the expected value cuts are initially present but are gradually replaced by stronger stochastic cuts.

Summary of Properties of Cuts for Inter-Stage Independence

Different scenarios ω_t in stage t are distinguished by different right-hand sides of the original stage t constraints, e.g., $A_t x_t = b_t^{\omega_t} + B_{t-1}^{\omega_t} \hat{x}_{t-1}$. The set of cuts $-G_t^{l_t} x_t + \theta_t \geq g_t^{l_t}, l_t = 1, \ldots, L_t$, represents an outer linearization of the expected future costs that is independent of stage t scenarios and is valid for all scenarios $\omega_t \in \Omega_t$. The outer linearization defined by the set of cuts equals the expected future cost function if $E z_{t+1}^{\omega_{t+1}}(\hat{x}_t) = \hat{\theta}_t$, where $\hat{\theta}_t$ is the value of θ_t corresponding to the solution $\hat{x}_t$ of any stage t problem. If $E z_{t+1}^{\omega_{t+1}}(\hat{x}_t^{\omega_t}) = \hat{\theta}_t^{\omega_t}$, for all $\omega_t \in \Omega_t$, then a sufficient number of necessary cuts has been generated to represent the expected future costs for all solutions $\hat{x}_t^{\omega_t}$ of scenarios $\omega_t \in \Omega_t$ in stage t, and we say that stage t is balanced with stage $t+1$. In any stage $t-1$, $t = 2, \ldots, T-1$, we can obtain cuts from stage t subproblems that are not balanced with their descendant stage $t+1$ subproblems. The resulting cuts are valid, weak at the x_{t-1} at which they are derived, but potentially efficient at some other value of

x_{t-1}. Cuts obtained in any stage $t-1$ from expected-value subproblems in stage t are valid for any scenario ω_{t-1}. They are weak, but they can be used to guide the algorithm at the beginning.

4.4.2 The Case of Dependency of Stochastic Parameters between Stages

In the case of inter-stage dependency of stochastic parameters, the coefficients and the right-hand sides of the cuts depend on the scenario history. Simple sharing of cuts between different scenario subproblems $\omega_t \in \Omega_t$ is no longer possible. However, for classes of linear (Markovian) dependency, cuts can be adjusted to fit different scenarios. (For example, Pereira and Pinto (1989) [103] discussed a class of additively dependent right-hand sides.)

In the auto-correlative model developed in Section 4.2.2, equations (4.8) and (4.9), the transition matrix and the right-hand side in stage t, based on the scenario history $\omega_{t-1}, \ldots, \omega_2$, are expressed as

$$B_{t-1}^{\omega_t,\ldots,\omega_2} = \mathrm{vec}_B^{-1}(\beta_0 v^{\omega_t}) + \sum_{\tau=1}^{t-2} \mathrm{vec}_B^{-1}(\beta_\tau \mathrm{vec}(B_{t-\tau-1}^{\omega_{t-\tau}}, b_{t-\tau}^{\omega_{t-\tau}})), \qquad (4.40)$$

$$b_t^{\omega_t,\ldots,\omega_2} = \mathrm{vec}_b^{-1}(\beta_0 v^{\omega_t}) + \sum_{\tau=1}^{t-2} \mathrm{vec}_b^{-1}(\beta_\tau \mathrm{vec}(B_{t-\tau-1}^{\omega_{t-\tau}}, b_{t-\tau}^{\omega_{t-\tau}})), \qquad (4.41)$$

where τ represents the index of the time lag.

In the case of inter-stage independence of the stochastic parameters, we compute a cut from the (dual) stage t problems,

$$z_t^{\omega_t}(\hat{x}_{t-1}) = \max \quad (\pi_t^{\omega_t}(b_t^{\omega_t} + B_{t-1}^{\omega_t}\hat{x}_{t-1}) + \rho_t^{\omega_t} g_t)$$
$$\mathrm{s/t}$$

$$\begin{aligned} \pi_t^{\omega_t} A_t \qquad - \rho_t^{\omega_t} G_t &\leq c_t \\ \rho_t^{\omega_t} 1 &= 1 \\ \rho_t^{\omega_t} &\geq 0, \; \omega_t \in \Omega_t, \end{aligned} \qquad (4.42)$$

as

$$\theta_{t-1} \geq E_{\omega_t}(\pi_t^{\omega_t}(b_t^{\omega_t} + B_{t-1}^{\omega_t} x_{t-1}) + \rho_t^{\omega_t} g_t). \tag{4.43}$$

In the case of inter-stage dependency of stochastic parameters, a cut takes on the form

$$\theta_{t-1} \geq E_{\omega_t}(\pi_t^{\omega_t,\ldots,\omega_2}(b_t^{\omega_t,\ldots,\omega_2} + B_{t-1}^{\omega_t,\ldots,\omega_2} x_{t-1}) + \rho_t^{\omega_t,\ldots,\omega_2} g_t^{\omega_t,\ldots,\omega_2}). \tag{4.44}$$

Note the scenario history $\omega_t, \ldots, \omega_2$ of the optimal dual variables $\pi_t^{\omega_t,\ldots,\omega_2}$ and $\rho_t^{\omega_t,\ldots,\omega_2}$. The history is induced through the gradients of the cuts in the stage t subproblem, as $G_t^{l_t}$ in the independent case becomes $G_t^{\omega_t,\ldots,\omega_2,l_t}$ in the dependent case. Including the inter-stage dependency according to the auto-correlative model by substitution leads to the following formulation:

$$\theta_{t-1} \geq$$

$$E_{\omega_t}\pi_t^{\omega_t,\ldots,\omega_2}(\mathrm{vec}_b^{-1}(\beta_0 v^{\omega_t})) + E_{\omega_t}\pi_t^{\omega_t,\ldots,\omega_2}(\mathrm{vec}_B^{-1}(\beta_0 v^{\omega_t})x_{t-1}) +$$

$$[(E_{\omega_t}\pi_t^{\omega_t,\ldots,\omega_2})(\sum_{\tau=1}^{t-2} \mathrm{vec}_b^{-1}(\beta_\tau \mathrm{vec}(B_{t-\tau-1}^{\omega_{t-\tau}}, b_{t-\tau}^{\omega_{t-\tau}}))) +$$

$$(E_{\omega_t}\pi_t^{\omega_t,\ldots,\omega_2})(\sum_{\tau=1}^{t-2} \mathrm{vec}_B^{-1}(\beta_\tau \mathrm{vec}(B_{t-\tau-1}^{\omega_{t-\tau}}, b_{t-\tau}^{\omega_{t-\tau}}))x_{t-1})] +$$

$$E_{\omega_t}(\rho_t^{\omega_t,\ldots,\omega_2} g_t^{\omega_t,\ldots,\omega_2}). \tag{4.45}$$

The obtained cut consists of three parts: the first part reflects the contribution of the outcomes of the stage t independent random vector $v_t^{\omega_t}$; the second part reflects the contribution of the observed outcomes in stages $2, \ldots, t-1$; and the third part concerns the update of the right-hand side coefficients. Still, a cut based on the auto-correlative dependency model cannot be shared directly between the subproblems in stage $t-1$. The optimal dual variables, $\pi_t^{\omega_t,\ldots,\omega_2}$ and $\rho_t^{\omega_t,\ldots,\omega_2}$, depend on ω_{t-1} and on the full history of realizations $\omega_{t-2}, \ldots, \omega_2$. The dependency is induced through the gradients of the cuts in the stage t problem, $G_t^{\omega_t,\ldots,\omega_2,l_t}$, which depend on $\omega_t, \ldots, \omega_2$. In order to enable sharing, cuts have to be valid; that is, the dual variables $\pi_t^{\omega_t,\ldots,\omega_2}$ and $\rho_t^{\omega_t,\ldots,\omega_2}$ have to be a feasible solution of the stage t, ω_t dual subproblem, for any history $\omega_{t-1},\ldots,\omega_2$. For the general case of the auto-correlative model stated above, establishing dual feasibility for all scenarios ω_t and any history $\omega_{t-1},\ldots,\omega_2$ is at best difficult. We do not discuss the general case of the auto-correlative model further. Instead, we focus the discussion on simpler cases, in which dual feasibility in each stage t is retained, independent of the scenario history up to stage t.

Inter-Stage Dependency of the Right-Hand Sides b_t and Inter-Stage Independence of the Transition Matrices B_{t-1}

For the case of inter-stage dependency of the right-hand sides b_t and inter-stage independence of the transition matrices B_{t-1}, we assume in the auto-correlative model that $\text{vec}_B^{-1}(\beta_\tau \text{vec}(B_{t-\tau-1}^{\omega_{t-\tau}}, b_{t-\tau}^{\omega_{t-\tau}})) = 0$ for all $\tau = 1,\ldots,t-2$ and consider the following model:

$$B_{t-1}^{\omega_t} = \text{vec}_B^{-1}(\beta_0 v^{\omega_t}), \tag{4.46}$$

$$b_t^{\omega_t,\ldots,\omega_2} = \text{vec}_b^{-1}(\beta_0 v^{\omega_t}) + \sum_{\tau=1}^{t-2} \text{vec}_b^{-1}(\beta_\tau \text{vec}(B_{t-\tau-1}^{\omega_{t-\tau}}, b_{t-\tau}^{\omega_{t-\tau}})). \tag{4.47}$$

We compute a cut from the (dual) stage t problem as

$$\theta_{t-1} \geq E_{\omega_t}(\pi_t^{\omega_t,\ldots,\omega_2}(b_t^{\omega_t,\ldots,\omega_2} + B_{t-1}^{\omega_t}x_{t-1}) + \rho_t^{\omega_t,\ldots,\omega_2}g_t^{\omega_t,\ldots,\omega_2}). \tag{4.48}$$

Including the inter-stage dependency according to the auto-correlative model by substitution leads to the following formulation:

$$\theta_{t-1} \geq$$
$$E_{\omega_t}\pi_t^{\omega_t,\ldots,\omega_2}(\text{vec}_b^{-1}(\beta_0 v^{\omega_t})) + E_{\omega_t}\pi_t^{\omega_t,\ldots,\omega_2}(\text{vec}_B^{-1}(\beta_0 v^{\omega_t})x_{t-1}) +$$
$$(E_{\omega_t}\pi_t^{\omega_t,\ldots,\omega_2})(\sum_{\tau=1}^{t-2}\text{vec}_b^{-1}(\beta_\tau \text{vec}(B_{t-\tau-1}^{\omega_{t-\tau}}, b_{t-\tau}^{\omega_{t-\tau}}))) +$$
$$E_{\omega_t}(\rho_t^{\omega_t,\ldots,\omega_2}g_t^{\omega_t,\ldots,\omega_2}). \tag{4.49}$$

Again, the obtained cut consists of three parts: the contribution of the outcomes of the stage t independent random vector $v_t^{\omega_t}$, the contribution of the observed outcomes in stages $2, \ldots, t-1$, and the update of the right-hand side coefficients. Note that, in case of inter-stage independence of the transition matrices $B_t^{\omega_t}$, the dual feasible region is independent of the scenario history $\omega_{t-1}, \ldots, \omega_2$. Thus, the optimal dual variables $\pi_t^{\omega_t, \ldots, \omega_2}$ and $\rho_t^{\omega_t, \ldots, \omega_2}$ are dual feasible for any history $\omega_{t-1}, \ldots, \omega_2$, and they therefore generate valid cuts. Moreover, any such cut can be adjusted to a given scenario $\omega_{t-1}, \ldots, \omega_2$ if the expectations $(E_{\omega_t} \pi_t^{\omega_t, \ldots, \omega_2})$ and $(E_{\omega_t} \rho_t^{\omega_t, \ldots, \omega_2})$ are known. We drop indices $\omega_{t-2}, \ldots, \omega_2$ because the property of dual feasibility is now independent of scenario history, and denote the dual variables only as $\pi_t^{\omega_t}$ and $\rho_t^{\omega_t}$. They are optimal for the particular scenario history $\omega_{t-2}, \ldots, \omega_2$ at which they have been generated, and feasible for any other scenario history. Now we can restate the cut as follows:

$$
\begin{aligned}
\theta_{t-1} \geq \\
E_{\omega_t} \pi_t^{\omega_t} (\mathrm{vec}_b^{-1}(\beta_0 v^{\omega_t})) + E_{\omega_t} \pi_t^{\omega_t} (\mathrm{vec}_B^{-1}(\beta_0 v^{\omega_t}) x_{t-1}) + \\
(E_{\omega_t} \pi_t^{\omega_t})(\textstyle\sum_{\tau=1}^{t-2} \mathrm{vec}_b^{-1}(\beta_\tau \mathrm{vec}(B_{t-\tau-1}^{\omega_{t-\tau}}, b_{t-\tau}^{\omega_{t-\tau}}))) + \\
E_{\omega_t}(\rho_t^{\omega_t} g_t^{\omega_t, \ldots, \omega_2}).
\end{aligned}
\tag{4.50}
$$

Note the notation of $g_t^{\omega_t, \ldots, \omega_2}$ to indicate that the right-hand sides of the cuts from stage $t+1$ have a history because of $b_{t+1}^{\omega_{t+1}, \ldots, \omega_2}$. However, based on the linear dependency model we can compute the stage t impact depending on scenario ω_t and the history $\omega_{t+1}, \ldots, \omega_2$. We do not discuss this in further detail here.

In order to adjust the cut to a given scenario, we need only specially compute the history-dependent part of the cut formulation. Given a scenario history $\omega_{t-2}, \ldots, \omega_2$, we update cuts by specifically recomputing the ω_{t-1}-dependent terms. These terms appear in the second and third parts of the cut formulation. The first part remains the same for each scenario ω_{t-1} and the history $\omega_{t-2}, \ldots, \omega_2$.

Note that the expected-value computation is computationally expensive. Because the first part remains the same for each scenario $\omega_{t-1}, \ldots, \omega_2$, we store the calculation and adjust the cut for a given scenario by adding the second and third parts using $(E_{\omega_t} \pi_t^{\omega_t})$, the expected value of the dual prices already obtained and stored during the computation of the first part. We now can conclude:

For the auto-correlative dependency model, with inter-stage dependency of the right-hand sides b_t and inter-stage independence of the transition matrices B_{t-1}, cuts in any stage t can be adjusted to be valid in any scenario $\omega_t \in \Omega_t$.

We further exhibit the procedure by considering a simplified example of the auto-correlative model of (4.40) and (4.41). We consider uncertainty only in the right-hand sides and auto-correlative dependency with a time lag of 1. We further consider, in order to ease the presentation, instead of matrices $\beta_\tau, \tau = 0, 1$, scalars $\beta_\tau^b, \tau = 0, 1$:

$$b_t^{\omega_t} = \beta_0^b \eta_t^{\omega_t} + \beta_1^b b_{t-1}^{\omega_{t-1}}. \tag{4.51}$$

In this formulation η_t represents a vector of the dimension of b_t whose elements are functions of random parameters that are independent of the random parameters of all previous periods. The corresponding cut,

$$\theta_{t-1} \geq E_{\omega_t} \pi_t^{\omega_t} (\beta_0^b \eta_t^{\omega_t} + B_{t-1} x_{t-1}) + (E_{\omega_t} \pi_t^{\omega_t}) \beta_1^b b_{t-1}^{\omega_{t-1}} + E_{\omega_t} \rho_t^{\omega_t} g_t^{\omega_t, \ldots, \omega_2}, \tag{4.52}$$

needs only to be adjusted in the right-hand side by specializing the term $(E_{\omega_t} \pi_t^{\omega_t})$ $\times \beta_1^b b_{t-1}^{\omega_{t-1}}$ and the scenario ω_{t-1}-dependent term of $E_{\omega_t} \rho_t^{\omega_t} g_t^{\omega_t, \ldots, \omega_2}$ to the stage $t-1$, ω_{t-1} scenario subproblem we currently want to solve.

Inter-Stage Dependency in Three-Stage Models

We pointed out before that, if we want to be able to share cuts in stage $t - 1$ for different scenarios ω_{t-1}, given a history $\omega_{t-2}, \ldots, \omega_2$, the stage t dual variables $\pi_t^{\omega_t, \ldots, \omega_2}$ have to be feasible for any history $\omega_{t-2}, \ldots, \omega_2$. This property is satisfied in the case discussed above, in which the stochastic parameters of the transition matrices $B_{t-1}^{\omega_{t-1}}$ are inter-stage independent. Similarly, the property of dual feasibility is retained if we restrict the number of stages of the multi-stage model to $T = 3$. In this case, the stage 3, ω_3 subproblems do not contain any cuts (as there are no further stages); therefore, the stage 3 dual feasible region is not affected by any historical outcome ω_2. For the three-stage model we can then formulate the general auto-correlative model, with inter-stage dependency of the transition matrices as well as the right-hand side.

The formulation of the second-stage cuts takes on the form

$$\theta_2 \geq E_{\omega_3}(\pi_3^{\omega_3,\omega_2}(b_3^{\omega_3,\omega_2} + B_2^{\omega_3,\omega_2}x_2)), \qquad (4.53)$$

and substituting the auto-correlative model leads to

$$
\begin{aligned}
\theta_2 \geq& \\
&E_{\omega_3}\pi_3^{\omega_3,\omega_2}(\mathrm{vec}_b^{-1}(\beta_0 v^{\omega_3})) + E_{\omega_3}\pi_3^{\omega_3,\omega_2}(\mathrm{vec}_B^{-1}(\beta_0 v^{\omega_3})x_2) + \\
&[(E_{\omega_3}\pi_3^{\omega_3,\omega_2})(\mathrm{vec}_b^{-1}(\beta_1 \mathrm{vec}(B_1^{\omega_2}, b_2^{\omega_2}))) + \\
&(E_{\omega_3}\pi_3^{\omega_3,\omega_2})(\mathrm{vec}_B^{-1}(\beta_1 \mathrm{vec}(B_1^{\omega_2}, b_2^{\omega_2}))x_2)].
\end{aligned}
\qquad (4.54)
$$

As before, we drop index ω_2 because the property of dual feasibility in stage 3 is independent of scenario ω_2, denoting the dual variables only as $\pi_t^{\omega_3}$, and restate the cut as follows:

$$
\begin{aligned}
\theta_2 \geq& \\
&E_{\omega_3}\pi_3^{\omega_3}(\mathrm{vec}_b^{-1}(\beta_0 v^{\omega_3})) + E_{\omega_3}\pi_3^{\omega_3}(\mathrm{vec}_B^{-1}(\beta_0 v^{\omega_3})x_2) + \\
&[(E_{\omega_3}\pi_3^{\omega_3})(\mathrm{vec}_b^{-1}(\beta_1 \mathrm{vec}(B_1^{\omega_2}, b_2^{\omega_2}))) + \\
&(E_{\omega_3}\pi_3^{\omega_3})(\mathrm{vec}_B^{-1}(\beta_1 \mathrm{vec}(B_1^{\omega_2}, b_2^{\omega_2}))x_2)].
\end{aligned}
\qquad (4.55)
$$

Compared to the formulation for the general multi-stage problem, the third part of the cut formulation has vanished; two parts, the first depending on ω_3 and the second depending on ω_2, remain. The cut can be adjusted to the current scenario ω_2 if the expectation $(E_{\omega_3}\pi_t^{\omega_3})$ is known. In order to adjust the cut to a given scenario ω_2, we specially compute the second and ω_2-dependent part of the cut formulation. The first part remains the same for each scenario ω_2. We store the calculation and adjust the cut for a given scenario by adding the second part using $(E_{\omega_3}\pi_3^{\omega_3})$, and we update the gradients as well as the right-hand side of the cut.

For three-stage models with auto-correlative dependency, cuts in stage 2 can be adjusted to be valid in any scenario $\omega_2 \in \Omega_2$.

We further demonstrate how to update cuts in the second stage with a simplified example of the general auto-correlative model of equations (4.40) and (4.41). We assume a dependency structure in which uncertainty occurs only in the transition matrix and consider, in order to ease the presentation, instead of matrices $\beta_\tau, \tau = 0, 1$, scalars $\beta_\tau^B, \tau = 0, 1$:

$$B_2^{\omega_3} = \beta_0^B \tilde{\eta}_3^{\omega_3} + \beta_1^B B_1^{\omega_2}, \tag{4.56}$$

$$B_1^{\omega_2} = \beta_0^B \tilde{\eta}_2^{\omega_2}, \tag{4.57}$$

where for $t = 2, 3$, $\tilde{\eta}_t^{\omega_t}$ is a matrix (of the dimension of B_{t-1}) of independent random parameters whose elements are functions of random parameters that are independent of the random parameters of previous periods. The corresponding cut,

$$\theta_2 \geq E_{\omega_3}\pi_3^{\omega_3}(b_3 + \beta_0^B \tilde{\eta}_3 x_{t-1}) + (E_{\omega_3}\pi_t^{\omega_3})\beta_1^B B_1^{\omega_2} x_{t-1}, \tag{4.58}$$

needs to be adjusted in its coefficients by specializing the term $(E_{\omega_3}\pi_t^{\omega_3})\beta_1^B B_1^{\omega_2}$ to the stage 2, ω_2 scenario subproblem to be solved.

4.4.3 Summary of Properties of Cuts

Taking advantage of the above-stated properties, we actually need only store one subproblem per stage t. For different scenarios ω_t and different solutions $\hat{x}_{t-1}$

passed from the previous stage, we determine the right-hand side of the subproblem accordingly. Cuts generated by stage $t + 1$ subproblems are valid for all scenarios $\omega_t \in \Omega_t$ in the case of independence of the stochastic parameters between stages. In the case of auto-correlative inter-stage dependency, certain special cases, as described above, allow for "sharing" of valid cuts among scenarios ω_t through the reuse of common expected dual prices from stage $t + 1$ and adjusting applied cuts according to the scenario history up to stage t, $\omega_t, \dots, \omega_2$.

Future information is represented in the cuts that have been generated so far and can be used in any scenario $\omega_t \in \Omega_t$ independently of which scenario originated the cut.

4.5 PROBABILISTIC LOWER BOUNDS

4.5.1 Estimates of Expected Values

For calculating the expected values of the future costs, the gradients, and the right-hand sides of the cuts in each stage t, we use Monte Carlo importance sampling as discussed in Section 2.3.3 for two-stage stochastic linear programs. Employing Monte Carlo sampling techniques means not solving all problems $\omega_{t+1} \in \Omega_{t+1}$ but solving problems $\omega_{t+1} \in S_{t+1}$, where S_{t+1} is a subset of Ω_{t+1}. Instead of the exact expected values $z_{t+1}(\hat{x}_t)$, $G_t(\hat{x}_t)$, and $g_t(\hat{x}_t)$, we compute estimates $\bar{z}_{t+1}(\hat{x}_t)$, $\bar{G}_t(\hat{x}_t)$, and $\bar{g}_t(\hat{x}_t)$ using the importance sampling procedure. We also estimate the error in the estimation of $z_{t+1}(\hat{x}_t)$ by the variance

$$\sigma^2_{\bar{z}_{t+1}}(\hat{x}_t) := \text{var}(\bar{z}_{t+1}(\hat{x}_t)). \tag{4.59}$$

Thus, given a particular $\hat{x}_t$, we obtain from the importance sampling procedure an estimate of the mean of the stage $t+1$ costs and the associated error distribution. For sample sizes larger than 40 (see, for example, Davis and Rabinowitz (1984) [32]), one can assume that the error of the estimation is normally distributed. Therefore, we define the estimate of the expected stage $t + 1$ costs $\tilde{z}_{t+1}(\hat{x}_t)$ for given $\hat{x}_t$ to be a random parameter, normally distributed with mean $\bar{z}_{t+1}(\hat{x}_t)$ and variance $\sigma^2_{\bar{z}_{t+1}}$:

$$\tilde{z}_{t+1}(\hat{x}_t) := N(\bar{z}_{t+1}(\hat{x}_t), \sigma^2_{\bar{z}_{t+1}}(\hat{x}_t)). \tag{4.60}$$

A cut with estimated coefficients and right-hand sides differs from a cut obtained by computing the expected values of the coefficients and right-hand sides exactly. The outer linearization

$$L_t(\hat{x}_t, x_t) = G_t(\hat{x}_t)x_t + g_t(\hat{x}_t) \tag{4.61}$$

with respect to the universe case and

$$\bar{L}_t(\hat{x}_t, x_t) = \bar{G}_t(\hat{x}_t)x_t + \bar{g}_t(\hat{x}_t) \tag{4.62}$$

with respect to the estimation differ in the gradient and in the right-hand side. We denote as

$$\epsilon_{t,\hat{x}_t}(x_t) = L_t(\hat{x}_t, x_t) - \bar{L}_t(\hat{x}_t, x_t) \tag{4.63}$$

the difference between the true (universe) and the estimated value of the outer linearization. At $x_t = \hat{x}_t$, the value at which the cut was derived, $L_t(\hat{x}_t, \hat{x}_t) = z_{t+1}(\hat{x}_t)$ and $\bar{L}_t(\hat{x}_t, \hat{x}_t) = \bar{z}_{t+1}(\hat{x}_t)$. Thus, if a true cut obtained by solving the universe case is binding at the solution $x_t = \hat{x}_t$, the variable θ_t takes on the value

$$\theta_t = L_t(\hat{x}_t, \hat{x}_t) = z_{t+1}(\hat{x}_t). \tag{4.64}$$

In the case of using Monte Carlo sampling, we relate θ_t to the estimated value of the expected stage $t + 1$ costs at $\hat{x}_t$, $\bar{L}_t(\hat{x}_t, \hat{x}_t) = \bar{z}_{t+1}(\hat{x}_t)$, and correct for estimation error by adjusting the right-hand side. Thus we can state

$$\theta_t = \bar{L}_t(\hat{x}_t, \hat{x}_t) - \bar{z}_{t+1}(\hat{x}_t) + z_{t+1}(\hat{x}_t), \tag{4.65}$$

$$\theta_t = \bar{L}_t(\hat{x}_t, \hat{x}_t) + \epsilon_{t,\hat{x}_t}(\hat{x}_t). \tag{4.66}$$

Equation (4.65) represents a valid statement for a solution $x_t = \hat{x}_t$. The correction term $\epsilon_{t,\hat{x}_t}(\hat{x}_t) = z_{t+1}(\hat{x}_t) - \bar{z}_{t+1}(\hat{x}_t)$ corrects for the estimation error. Of course, we do not know the difference $z_{t+1}(\hat{x}_t) - \bar{z}_{t+1}(\hat{x}_t)$ explicitly for each cut when we compute it. However, we can obtain an estimate of the distribution of the correction term from the estimation process. Recall that by using Monte Carlo importance sampling, we obtain an unbiased estimate of $z_{t+1}(\hat{x}_t)$, $\bar{z}_{t+1}(\hat{x}_t)$, with variance $\sigma^2_{\bar{z}_{t+1}}(\hat{x}_t)$. Therefore, $\epsilon_{t,\hat{x}_t}(\hat{x}_t)$ is normally distributed with mean 0 and variance $\sigma^2_{\bar{z}_{t+1}}(\hat{x}_t)$:

$$\epsilon_{t,\hat{x}_t}(\hat{x}_t) := N(0, \sigma^2_{\bar{z}_{t+1}}(\hat{x}_t)). \qquad (4.67)$$

Suppose that a cut $\bar{L}_t(\hat{x}_t, x_t) = \bar{G}_t(\hat{x})x + \bar{g}(\hat{x})$, computed at $x_t = \hat{x}_t$, is binding at a solution $\hat{\hat{x}}_t$ where $\hat{x}_t \neq \hat{\hat{x}}_t$. Applying again a correction for the estimation error, we obtain:

$$\theta_t = \bar{L}_t(\hat{x}_t, \hat{\hat{x}}_t) + \epsilon_{t,\hat{x}_t}(\hat{\hat{x}}_t). \qquad (4.68)$$

The correction term for the estimation error is clearly the true value of the cut at $x_t = \hat{\hat{x}}_t$ minus the value obtained by the sampling procedure:

$$\epsilon_{t,\hat{x}_t}(\hat{\hat{x}}) = L_t(\hat{x}_t, \hat{\hat{x}}) - \bar{L}_t(\hat{x}_t, \hat{\hat{x}}). \qquad (4.69)$$

Again we do not know the difference $L_t(\hat{x}_t, \hat{\hat{x}}) - \bar{L}_t(\hat{x}_t, \hat{\hat{x}})$ when we compute the cut. The distribution of the estimation error, $\epsilon_{t,\hat{x}_t}(\hat{\hat{x}})$ at $x_t = \hat{x}_t$, can be computed based on the observations $G_t^{\omega t}$ and $g_t^{\omega t}$, which have been obtained by computing the estimation error of the value of a cut as a function of x.

For most practical problems it is a sufficiently close approximation to assume that

$$\epsilon_{t,\hat{x}_t}(\hat{\hat{x}}_t) \approx \epsilon_{t,\hat{x}_t}(\hat{x}_t) \quad \text{for} \quad \hat{\hat{x}}_t \approx \hat{x}_t. \qquad (4.70)$$

This means that we assume the error distribution $\epsilon_{t,\hat{x}_t}(x_t)$ to be constant with respect to x_t, and use $\epsilon_{t,\hat{x}_t}(\hat{x}_t)$, rather than taking into account correctly $\epsilon_{t,\hat{x}_t}(x_t)$ as a function of x_t.

However, everything that will be derived using the constant error term $\epsilon_{t,\hat{x}_t}(\hat{x}_t)$ can be extended to use the variable error term $\epsilon_{t,\hat{x}_t}(x_t)$. We will show this at the relevant places in the text.

We will in the following denote the constant error term $\epsilon_{t,\hat{x}_t}(\hat{x}_t)$ as ϵ_t or as $\epsilon_t^{l_t}$, short for $\epsilon_{t,\hat{x}_t^{l_t}}(\hat{x}_t^{l_t})$, when we refer to it in a particular iteration l_t of the Benders decomposition algorithm. We will denote the error term as a function of x_t, $\epsilon_{t,\hat{x}_t}(x_t)$ as $\epsilon_t(x_t)$ or as $\epsilon_t^{l_t}(x_t)$, short for $\epsilon_{t,\hat{x}_t^{l_t}}(x_t)$, when referring to it in a particular Benders iteration l.

4.5.2 The Lower Bound Estimate

In each stage t, $t = 2, \ldots, T - 1$, a lower bound for the stage t expected costs (subject to scenario ω_t and the solution $\hat{x}_{t-1}$ passed from the previous stage) is represented by the optimal objective function value of the stage t master problem, $z_t^{\omega_t}$. In each stage t, $t = 2, \ldots, T - 1$, the cuts added so far represent an outer linearization of the expected future costs. For an optimal solution $(\hat{x}_t, \hat{\theta}_t)$ of the stage t master problem, $\hat{\theta}_t$ represents a lower bound for the value of the expected future costs. A lower bound for the total expected costs of the multi-stage stochastic linear program is represented by the optimal objective function value of the stage 1 master problem, z_1, which includes a value for $\hat{\theta}_1$ representing a lower bound for the value of the expected future costs (in stages $2, \ldots, T$).

If cuts have been obtained in stage t by solving the stage $t + 1$, ω_{t+1} sub-problems, where stage $t + 1$ is not balanced with stage $t + 2$ (i.e., the stage $t + 1$ cuts do not fully represent the stage $t + 2$ expected costs), the stage t cuts are potentially weak cuts. They are not supports of the recourse function at the x_t from which they were derived, but they have the potential to be supports of the recourse function at some other x_t. This situation is the case if, in stage $t + 1$, for a solution $\hat{x}_{t+1}^{\omega_{t+1}}, \hat{\theta}_{t+1}^{\omega_{t+1}}$,

$$\hat{\theta}_{t+1}^{\omega_{t+1}} < z_{t+2}(\hat{x}_{t+1}^{\omega_{t+1}}). \tag{4.71}$$

The corresponding lower bound is then a weak lower bound.

If the cuts are obtained by Monte Carlo sampling rather than by solving the universe case, the optimal objective function value of the stage t, ω_t problem, $z_t^{\omega_t}$, represents an **estimated** lower bound for the stage t, ω_t expected costs (subject to $\hat{x}_{t-1}$ passed from the previous stage). The optimal objective function value of the stage 1 problem, z_1, represents an estimate for the lower bound of the expected costs of the multi-stage stochastic linear program.

In the following we derive a lower bound estimate for the multi-stage stochastic linear problem by analyzing the decomposed program, where at stages $t = 1, \ldots, T-1$, L_t cuts estimated by Monte Carlo (importance) sampling have been added. We start with an analysis of the relationship between stage $T-1$ and stage T.

The stage $T-1$ master problem:

$$
\begin{aligned}
\tilde{z}_{T-1}^{\omega_{T-1}} \;=\; & \min \; c_{T-1} x_{T-1}^{\omega_{T-1}} \;+\; \theta_{T-1}^{\omega_{T-1}} \\
& \text{s/t}
\end{aligned}
$$

$$
\begin{aligned}
\pi_{T-1}^{\omega_{T-1}} : \quad & A_{T-1} x_{T-1}^{\omega_{T-1}} && = b_{T-1}^{\omega_{T-1}} + B_{t-2}^{\omega_{T-1}} \hat{x}_t \\
\rho_{T-1}^{1,\omega_{T-1}} : \quad & -\bar{G}_{T-1}^1 x_{T-1}^{\omega_{T-1}} + \theta_{T-1}^{\omega_{T-1}} && \geq \bar{g}_{T-1}^1 + \epsilon_{T-1}^1 \\
& \qquad\qquad\qquad \vdots && \vdots \\
\rho_{T-1}^{L_{T-1},\omega_{T-1}} : \quad & -\bar{G}_{T-1}^{L_{T-1}} x_{T-1}^{\omega_{T-1}} + \theta_{T-1}^{\omega_{T-1}} && \geq \bar{g}_{T-1}^{L_{T-1}} + \epsilon_{T-1}^{L_{T-1}} \\
& \qquad\qquad\qquad x_{T-1}^{\omega_{T-1}} && \geq 0.
\end{aligned} \tag{4.72}
$$

The stage T subproblem:

$$
\begin{aligned}
\min \; z_T^{\omega_T} \;=\; & c_T x_T^{\omega_T} \\
& \text{s/t} \\
\pi_T^{\omega_T} : \quad & A_T x_T^{\omega_T} = b_T^{\omega_T} + B_{T-1}^{\omega_T} \hat{x}_{T-1} \\
& x_T^{\omega_T} \geq 0.
\end{aligned} \tag{4.73}
$$

For a given scenario ω_{T-1} and a given solution $\hat{x}_{T-2}$ passed from the stage $T-1$ problem, the relation between stage $T-1$ and stage T corresponds to that of a two-stage stochastic linear program. We can therefore apply the theory we have developed for solving two-stage stochastic linear programs directly.

In the stage $T - 1$ problem, L_{T-1} cuts have been added so far. These cuts have been obtained by passing solutions $\hat{x}_{T-1}^{l_{T-1}}$, $l_{T-1} = 1, \ldots, L_{T-1}$, to stage T and solving a sample set $S_T^{l_{T-1}}$ of stage T subproblems according to the importance sampling scheme. By doing so, we have obtained estimates of the stage T expected costs $z_T(\hat{x}_{T-1}^{l_{T-1}})$ and of the gradients $G_{T-1}^{l_{T-1}}$ and right-hand sides $g_{T-1}^{l_{T-1}}$ of the cuts. These estimates are denoted by $\bar{z}_T(\hat{x}_{T-1}^{l_{T-1}})$, $\bar{G}_{T-1}^{l_{T-1}}$ and $\bar{g}_{T-1}^{l_{T-1}}$. We have also obtained the error correction terms $\epsilon_{T-1}^{l_{T-1}}(x_{T-1})$ based on the estimation error at $\hat{x}_{T-1}^{l_{T-1}}$, which for ease of exposition we assume to be constant with respect to x_t. In general their distributions vary with respect to x_{T-1}; see Section 2.4.2. However, assuming the error is constant with respect to x_{T-1}, each error term $\epsilon_{T-1}^{l_{T-1}}$ then arises from a normal distribution with mean 0 and variance $\text{var}(\bar{z}_T(\hat{x}_{T-1}^{l_{T-1}}))$. We denote

$$(\sigma_{T-1}^{l_{T-1}})^2 := \text{var}(\bar{z}_T(\hat{x}_{T-1}^{l_{T-1}})). \tag{4.74}$$

Thus,

$$\epsilon_{T-1}^{l_{T-1}} = N(0, (\sigma_{T-1}^{l_{T-1}})^2). \tag{4.75}$$

Based on a local error analysis of the stage $T - 1$ problem (4.72) in scenario ω_{T-1} given $\hat{x}_{T-2}$, the optimal objective function value $\bar{z}_{T-1}^{\omega_{T-1}}$ is a random parameter, normally distributed with mean $\bar{z}_{T-1}^{\omega_{T-1}}$ and variance $\text{var}(\bar{z}_{T-1}^{\omega_{T-1}})$:

$$\tilde{z}_{T-1}^{\omega_{T-1}} = N(\bar{z}_{T-1}^{\omega_{T-1}}, \text{var}(\bar{z}_{T-1}^{\omega_{T-1}})), \tag{4.76}$$

where

$$(\sigma_{T-1}^{\omega_{T-1}})^2 := \text{var}(\bar{z}_{T-1}^{\omega_{T-1}}) = \sum_{l_{T-1}=1}^{L_{T-1}} (\rho_{T-1}^{\omega_{T-1}, l_{T-1}})^2 (\sigma_{T-1}^{l_{T-1}})^2. \tag{4.77}$$

The distribution of the stage $T-1$, scenario ω_{T-1} expected costs, $\tilde{z}_{T-1}^{\omega_{T-1}}$, is induced by the estimation errors in the cuts that have been obtained so far in stage $T-1$. The distribution is different for different scenarios ω_{T-1} because of the different dual variables $\rho_{T-1}^{\omega_{T-1}, l_{T-1}}$.

Next we discuss the general case of the relation between stage $t-1$ and stage t, $t = 2, \ldots, T-1$. In this case, the subproblems in stage t contain cuts that have been computed by Monte Carlo importance sampling and therefore represent an estimate of the outer linearization of the future stage $t+1$ expected costs.

The stage t subproblem:

$$
\begin{aligned}
\tilde{z}_t^{\omega_t} \quad &= \quad \min \ c_t x_t^{\omega_t} \ + \ \theta_t^{\omega_t} \\
&\text{s/t}
\end{aligned}
$$

$$
\begin{aligned}
\pi_t^{\omega_t} : \qquad & A_t x_t^{\omega_t} && = \ b_t^{\omega_t} + B_{t-1}^{\omega_t} \hat{x}_{t-1} \\
\rho_t^{1,\omega_t} : \qquad & -\bar{G}_t^1 x_t^{\omega_t} \ + \ \theta_t^{\omega_t} && \geq \ \bar{g}_t^1 + \epsilon_t^1 \\
& \quad \vdots \qquad\qquad \vdots \qquad \vdots \\
\rho_t^{L_t,\omega_t} : \qquad & -\bar{G}_t^{L_t} x_t^{\omega_t} \ + \ \theta_t^{\omega_t} && \geq \ \bar{g}_t^{L_t} + \epsilon_t^{L_t} \\
& \qquad\qquad\quad x_t^{\omega_t} && \geq \ 0.
\end{aligned}
\qquad (4.78)
$$

In the analysis of the stage $T-1$ master problem we concluded that the objective value in stage $T-1$, scenario ω_{T-1}, $\tilde{z}_{T-1}^{\omega_{T-1}}$, is normally distributed with mean $\bar{z}_{T-1}^{\omega_{T-1}}$ and variance $(\sigma_{T-1}^{\omega_{T-1}})^2 = \mathrm{var}(\tilde{z}_{T-1}^{\omega_{T-1}})$. In the following we show that for any stage t, $t = 2, \ldots, T-2$, the expected costs $\tilde{z}_t^{\omega_t}$ are normally distributed with mean $\bar{z}_t^{\omega_t}$ and variance $\mathrm{var}(\bar{z}_t^{\omega_t})$, which we will denote by $(\sigma_t^{\omega_t})^2$:

$$
\tilde{z}_t^{\omega_t} = N(\bar{z}_t^{\omega_t}, (\sigma_t^{\omega_t})^2), \qquad (4.79)
$$

where

$$
(\sigma_t^{\omega_t})^2 := \mathrm{var}(\bar{z}_t^{\omega_t}) = \sum_{l_t=1}^{L_t} (\rho_t^{\omega_t, l_t})^2 (\sigma_t^{l_t})^2 \qquad (4.80)
$$

and

$$(\sigma_t^{l_t})^2 := \text{var}(\tilde{z}_{t+1}(\hat{x}_t^{l_t})). \qquad (4.81)$$

In order to do so, we discuss the relation between stage $t-1$ and stage t, where the stage t, ω_t problems are subproblems of the stage $t-1$, ω_{t-1} master problem. Given a solution $\hat{x}_{t-1}$ passed from the current stage $t-1$, ω_{t-1} master problem, the stage t costs, $\tilde{z}_t$, are distributed both with respect to ω_t and with respect to the error distributions ϵ_t that appear at the right-hand sides of the cuts:

$$\tilde{z}_t = \tilde{z}_t(\omega_t, \epsilon_t^1, \ldots, \epsilon_t^{L_t}). \qquad (4.82)$$

The expected value of the stage t costs with respect to the ω distribution, $\tilde{z}_t$, can be computed as

$$\tilde{z}_t = E_{\omega_t}(\tilde{z}_t \mid \omega_t) = E_{\omega_t}(\tilde{z}_t^{\omega_t}) \qquad (4.83)$$

using Monte Carlo importance sampling. Remember that $\tilde{z}_t^{\omega_t} = N(\tilde{z}_t^{\omega_t}, \text{var}(\tilde{z}_t^{\omega_t}))$. We compute the mean value $\tilde{z}_t^{\omega_t}$ by substituting the mean value of 0 of the error distributions $\epsilon_t^{l_t}$. That is, the current estimate of the mean is the current objective value.

We compute the variance $\text{var}(\tilde{z}_t)$ based on conditional expectations:

$$
\begin{aligned}
\text{var}(\tilde{z}_t) &= E_{\omega_t}\left(\text{var}(\tilde{z}_t) \mid \omega_t\right) + \text{var}_{\omega_t}(E_{\epsilon_t}\tilde{z}_t \mid \omega_t) \\
&= E_{\omega_t}\text{var}(\tilde{z}_t^{\omega_t}) + \text{var}_{\omega_t}(\tilde{z}_t^{\omega_t})
\end{aligned}
\qquad (4.84)
$$

$$
\begin{array}{ccc}
\text{var}(\tilde{z}_t) &= E_{\omega_t}\text{var}(\tilde{z}_t^{\omega_t}) &+ \frac{1}{n}\text{var}_{\omega_t}\tilde{z}_t^{\omega_t} \\
&(\sigma_{t,CUT})^2 &(\sigma_{t,SAMPL})^2 \\
&\textbf{[future]} &\textbf{[current]}.
\end{array}
\qquad (4.85)
$$

The variance of the estimated mean of the stage t costs can be seen as being composed of a sum of two terms. The first term concerns the influence of the error of the estimated cuts. It is an aggregate of the estimation error of all future periods $t+2, \ldots, T$. We refer to it as the **future** term of the estimation error. For constant error distributions, $\text{var}(\bar{z}_t^{\omega_t}) = \sum_{l_t=1}^{L_t} (\rho_t^{\omega_t, l_t})^2 (\sigma_t^{l_t})^2$. When we consider the estimation error as dependent on $\hat{x}_t$, then $\text{var}(\bar{z}_t^{\omega_t}) = \sum_{l_t=1}^{L_t} (\rho_t^{\omega_t, l_t})^2 (\sigma_t^{l_t}(\hat{x}_t))^2$. The second term concerns the **current** estimation error due to using sampling in stage t. In the case of two-stage problems, only the second term appears because the subproblems do not contain estimated cuts representing future information. Given $\hat{x}_{t-1}$ passed from the stage $t-1$ master problem, we can define the variance of the estimate $\text{var}(\bar{z}_t)$ to be

$$(\sigma_{t-1}^{l_{t-1}})^2 := \text{var}(\bar{z}_t(\hat{x}_{t-1}^{l_{t-1}})). \tag{4.86}$$

Thus, the expected value of the stage t costs is normally distributed with mean $\bar{z}_t(\hat{x}_{t-1}^{l_{t-1}})$ and variance $(\sigma_{t-1}^{l_{t-1}})^2$:

$$\tilde{z}_t(\hat{x}_{t-1}^{l_{t-1}}) := N(\bar{z}_t(\hat{x}_{t-1}^{l_{t-1}}), (\sigma_{t-1}^{l_{t-1}})^2). \tag{4.87}$$

The error correction terms of the stage $t-1$ cuts are represented as

$$\epsilon_{t-1}^{l_{t-1}} = N(0, (\sigma_{t-1}^{l_{t-1}})^2), \tag{4.88}$$

where we assume that the error distribution is approximately constant with respect to x_{t-1}. The minimum costs in stage $t-1$, scenario ω_{t-1}, $z_{t-1}^{\omega_{t-1}}$ are normally distributed with mean $\bar{z}_{t-1}^{\omega_{t-1}}$ and variance $(\sigma_{t-1}^{\omega_{t-1}})^2$:

$$\tilde{z}_{t-1}^{\omega_{t-1}} = N(\bar{z}_{t-1}^{\omega_{t-1}}, (\sigma_{t-1}^{\omega_{t-1}})^2). \tag{4.89}$$

Knowing the distribution of the error terms of the cuts in each stage t, $t = 2, \ldots, T-1$, we can compute an estimate of the mean of the stage t costs and the corresponding estimated variance of the mean value of the costs.

At stage 1 we obtain an estimate $\bar{z}_1$ with corresponding variance $\mathrm{var}(\bar{z}_1)$, where we define

$$(\sigma_1)^2 := \mathrm{var}(\bar{z}_1), \tag{4.90}$$

and

$$(\sigma_1)^2 = \sum_{l_1=1}^{L_1} (\rho_1^{l_1})^2 (\sigma_1^{l_1})^2. \tag{4.91}$$

The distribution of the optimal first-stage costs,

$$\tilde{z}_1 = N(\bar{z}_1, (\sigma_1)^2), \tag{4.92}$$

represents a lower bound estimate for the expected costs of the multi-stage stochastic linear program.

4.5.3   The Upper Bound Estimate

To obtain an upper bound for the total expected costs of the multi-stage problem, we evaluate the expected costs of the current first-stage trial solution $\hat{x}_1$. This can be accomplished by sampling paths from stages $2, \ldots, T$. For reference, see Pereira and Pinto (1989) [103]. To efficiently sample a small number of paths so as to obtain an accurate estimate of the expected costs associated with $\hat{x}_1$, we also use importance sampling. We define a path $\hat{s}^\omega = (\hat{x}_1, \hat{x}_2, \ldots, \hat{x}_T)^\omega$, $\omega \in \Omega$, where $\Omega = \{\Omega_2 \times \Omega_3 \times \cdots \times \Omega_T\}$, as a sequence of optimal solutions $\hat{x}_t^{\omega_t}$ of stage t, scenario ω_t problems, $t = 2, \ldots, T$, with $\hat{x}_1$ being the first-stage trial solution. A path is computed by following the "wait and see" paradigm. We pass $\hat{x}_1$ to the second stage and solve the second-stage problem for scenario ω_2 to obtain the optimal solution $\hat{x}_2^{\omega_2}$. Next we pass the obtained second-stage solution $\hat{x}_2^{\omega_2}$ to the third stage and solve the third-stage problem for scenario ω_3 to obtain $\hat{x}_3^{\omega_3}$. We continue in this way until we obtain $\hat{x}_T^{\omega_T}$ in stage T. Note that when solving the stage t problem, no future outcomes $\omega_{t+1}, \ldots, \omega_T$ are used. All future information at each stage is represented solely by means of the cuts added in stage t so far. The costs of a path $\hat{s}^\omega$, $C(\hat{s}^\omega)$, is given by $C(\hat{s}^\omega) = \sum_{t=1}^{T} c_t \hat{x}_t^{\omega_t}$. The expected value of the costs of all paths $\hat{s}^\omega$, $E\,\hat{s}^\omega$, gives an upper bound for the costs of a trial solution $\hat{x}_1$. Figure 4.2 represents schematically five paths sampled through the four-stage problem of Figure 4.1.

We sample paths by applying the importance sampling scheme to the space of dimension $\sum_{t=2}^{T} h_t$ of all random parameters $V_{i_t,t}$, $i_t = 1, \ldots, h_t$, $t = 2, \ldots, T$. For sampling paths the importance density $q(V)$ is computed based on the additive marginal approximation function $\Gamma(V)$ analogous to the way it was defined in

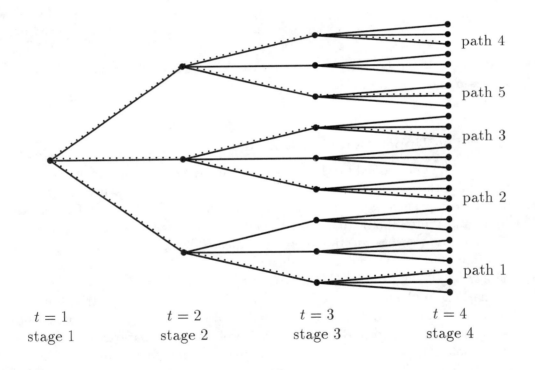

Figure 4.2: Path sampling for upper bounds

Section 2.3.3:

$$\Gamma(V) = C(\tau) + \sum_{t=1}^{T} \sum_{i_t=1}^{h_t} C(\tau_{1,1}, \ldots, \tau_{i_t-1,t}, V_{i_t,t}, \tau_{i_t+1,t}, \ldots, \tau_{h_T,T}) - C(\tau), \quad (4.93)$$

where $V = (V_{1,1}, \ldots, V_{h_1,1}, V_{1,2}, \ldots, V_{h_T,T})$ and $\tau = (\tau_{1,1}, \ldots, \tau_{h_1,1}, \tau_{1,2}, \ldots, \tau_{h_T,T})$. Using importance sampling for the upper bound estimate, we hope to obtain accurate estimates with a small sample size. Note that the advantage of sampling paths lies in the fact that we only linearly increase the number of sample points with the number of stages, whereas the decision tree grows exponentially with the number of stages.

4.6   THE ALGORITHM

By solving a sample of subproblems ω_{t+1} according to the importance sampling scheme, we compute estimates of the expected future costs $z_{t+1}^{\omega_t}$ and of the gradients $G_t^{l_t}$ and right-hand sides $g_t^{l_t}$ of the cuts in each stage t and scenario ω_t. The optimal objective function value for each stage t, scenario ω_t subproblem gives an estimated lower bound for the expected costs $z_t^{\omega_t} = c_t \hat{x}_t^{\omega_t} + \hat{\theta}_t^{\omega_t}$, subject to scenario ω_t and subject to $\hat{x}_{t-1}$, the (optimal) solution passed forward from the previous stage. The obtained lower bound estimate is the tightest lower bound that can be generated, if in stage $t+1$ a sufficient number of cuts have been added to represent the expected future costs with respect to stage $t+2$ for all scenarios $\omega_{t+1} \in \Omega_{t+1}$; it is a weaker lower bound estimate if there is not a sufficient number of cuts.

We are especially interested in the lower bound estimate of the first-stage costs, which we obtain by solving the first-stage problem. If the first-stage problem is balanced with the second stage (that is, if the cuts added so far to the first-stage problem fully represent the expected second-stage costs), and if the second stage is balanced with the third stage for all scenarios $\omega_2 \in \Omega_2$ and all values of $\hat{x}_1$, and so forth until stage $T - 1$, then the solution of the first-stage problem is the optimum solution of the multi-stage stochastic linear program. In this case, the lower bound estimate of z_1 takes on the value of the total expected costs of the multi-stage problem.

To estimate an upper bound for the total expected costs of the multi-stage problem, we use the path-sampling scheme with importance sampling to evaluate the expected costs of the current first-stage trial decision $\hat{x}_1$. Sampling paths $\omega \in \Omega$ according to this importance sampling scheme, we obtain an equal number of sample points $\omega_t \in \Omega_t$ in stages $t = 2, \ldots, T$. Figure 4.3 represents schematically these sample points for the example of the five paths of Figure 4.2. At these sample points we define the current stage t scenario ω_t subproblems and generate cuts to be added at stages $t = 1, \ldots, T - 1$, by employing importance sampling as described above for cuts.

The overall procedure works as follows. Solving the stage 1 problem at iteration 1, we obtain a trial solution $\hat{x}_1$ and a lower bound estimate for the expected costs z_1. Now we employ the path-sampling procedure to obtain an upper bound estimate for the expected costs z_1. If the upper bound estimate and the lower bound estimate are within a given optimality tolerance, we call the first-stage solution the optimal solution of the multi-stage problem and quit. Otherwise, we generate cuts in stages $1, \ldots, T - 1$. The path-sampling procedure used for the upper bound estimate has produced sample points $\omega_t \in \Omega_t$ in stages $t = 2, \ldots, T$, with corresponding ancestor solutions $\hat{x}_1$ and $\hat{x}_t^{\omega_t}$ in stages $t = 2, \ldots, T - 1$, to be passed to the current stage t scenario ω_t problem. Starting at stage $T - 1$ and moving backward to stage 1, we take each sample problem ω_t in stage t and finally the stage 1 problem as the current master problem and compute cuts by again sampling $\omega_{t+1} \in \Omega_{t+1}$ descendant subproblems until each scenario problem

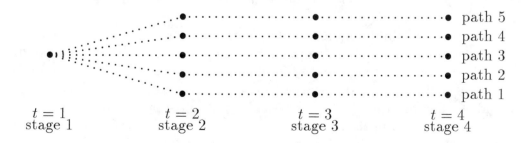

Figure 4.3: Sample points obtained from path sampling

ω_t in stage t is balanced with stage $t+1$ with respect to ancestor solutions $\hat{x}_{t-1}$, which have been passed from stage $t-1$. Arriving at stage 1, we obtain a new solution $\hat{x}_1$ and a new lower bound estimate. We continue as defined above by sampling new paths for the upper bound estimate. Finally, after a finite number of iterations, upper and lower bound estimates will be sufficiently close. Upper and lower bound estimates can be seen as sums of *i.i.d.* random terms, which for sample sizes of 30 or more can be assumed normally distributed with known variances (derived from the sampling process). A 95% confidence interval for the obtained solution is computed.

4.7 COMPUTATIONAL EXPERIENCE

Computational results from using Benders decomposition and importance sampling for two-stage problems and a special class of multi-stage problems are discussed in Section 2.7, where we report on the solution of test problems with up to 52 stochastic parameters and a number of universe scenarios exceeding 10^{24}. Using importance sampling and small sample sizes (between 200 and 600), we obtained very accurate results. Additional tests on these examples showed that the variance reduction factor obtained by using importance sampling versus crude (naive) Monte Carlo sampling was up to about 10^{-6}.

Inspired by these results, we implemented an earlier version of the methodology described above for the multi-stage case that did not consider dependency between stages. Instead of the path-sampling procedure for obtaining upper bound estimates, we implemented a procedure of sampling points that required handling of exponentially growing decision trees. Therefore, even when we used very small sample sizes, the number of stages that was practical to solve was limited.

We did test up to three-stage problems. FI3 is a three-stage test problem derived from a two-stage financial portfolio problem found in Mulvey and Vladimirou (1989) [96]. The problem is to select a portfolio that maximizes expected returns in future periods, taking into account the possibility of revising the portfolio in each period. There are transaction costs and bounds on the holdings and turnovers. Our test problem covers a planning horizon of three periods, whereas the original Mulvey-Vladimirou test problem was a two-stage problem that compressed all future periods into a single second stage. They solved the stochastic problem by restricting the number of scenarios.

We assumed the returns of the stocks in the future periods to be independent stochastic parameters with three outcomes each. With 13 assets with uncertain returns, the problem had 26 stochastic parameters instead of 39, because after the last-stage decision is made, the expected money value of the portfolio can be evaluated. The number of universe scenarios was $2.5 \cdot 10^{12}$. (The deterministic equivalent formulation of the problem would have more than 10^{14} rows and a similar number of columns.) We obtained an estimated optimal solution of the three-stage stochastic problem using a sample size of only 50 per stage. The optimal objective function value was estimated to be 1.10895 with an estimated 95% confidence interval of 0.004% on the left side and 0.001% on the right side of the obtained objective function value. Thus the optimal objective value lies within $1.10881 \leq z^* \leq 1.10895$ with 95% probability. Note how small the confidence interval is.

4.8 APPLICATIONS OF MULTI-STAGE STOCHASTIC LINEAR PROGRAMS

In the following we discuss two potential applications of multi-stage stochastic linear programs, one in the area of operations planning of power systems and the other in the area of portfolio optimization.

4.8.1 The Control of Hydro Power Systems

An important problem is the control of the short-term operation of hydro power systems. A river system can be described as a network of river basins, where outflows from one basin are inflows to another basin down the river. Powerhouses with different numbers of turbine-generator blocks operating in parallel for producing electrical energy are associated with each of the basins. There are two different types of outflows from basins: powerhouse (or turbine) flows and spill flows. The former go through the turbines and are used for energy generation; the latter bypass the turbines and cannot be used for energy generation. Spill is not desired in regular operation and is used only in situations when basin capacities and powerhouse-flow capacities are reached. Inflows into basins either result from outflows from basins up the river or are additional exogenous inflows. The exogenous inflows include all natural inflows into the river system (for example, rivers not used for power generation) and inflows resulting from reservoirs operated independently of the system under consideration.

The operating environment can be stated as follows. The river system is operated by a local load dispatcher as an independent economic entity. Energy produced by the river system is transmitted into the high-voltage system of a higher-level power company (producer, transmitter, and distributor) operated by the principal load dispatcher. The local load dispatcher receives requests for energy production from the principal load dispatcher. Prices paid depend on marginal cost calculations and vary with season and time of day. The local load dispatcher has to determine a schedule of operation for the river system that fulfills the requests of the principal load dispatcher, complies with various restrictions on the operation, and is based on projections of exogenous inflows and planned operation of independently operated reservoirs whose outflows are inflows to the river system.

Longer-term operations planning is conducted based on optimizations of planning horizons of a week, a day, or a weekend. Results from these longer-term optimizations influence short-term control by defining start and end values for reservoir levels. In fact, the local load dispatcher determines a rough schedule of the operation of the system for the next day based on planned energy requests and projections of inflows. Based on this rough daily schedule, on actual changes of the planned requests, and on short-term hydrological projections of inflows, the operation of the system has to be controlled for the immediate future. This means controlling actual operation of the system for the next four hours.

Predictions of natural inflows are usually based on the mean value of inflows of the previous day, the actually measured current value of inflow, and the hydrological forecast. Although hydrological forecasts are very inaccurate when extended to several days ahead, they can be reasonably good for shorter horizons— for example, the next day. However, there is still important uncertainty involved because the hydrological forecasts depend heavily upon local weather conditions, which cannot usually be predicted with sufficient accuracy. Inflows resulting from the operation of independently operated reservoirs are also subject to uncertainty because changes in electricity demand may result in a different schedule from the one predicted beforehand.

The control of the system has to have the ability to hedge against the stochastic nature of water availability. It has to protect against situations of shortage so that a sudden lack of inflow does not result in being unable to meet energy demand requests. It also has to be able to take advantage of situations of surplus inflow, so that a sudden, unforeseen increase in inflow does not result in having to spill water without being able to use it to produce electricity. It is clear that the system has to be able to adapt. That means it has to be able to observe when inflows differ from those projected and be able to take action accordingly.

The short-term control of the operation of the system continuously determines the levels of all turbine flows and spill flows. Given turbine and spill flows and given actual observations of all exogenous flows, all other variables of the hydraulic system are determined—the volumes and levels of water in the reservoirs, for example. Also, given turbine and spill flows and given the characteristics of the turbines and generators, the electric power produced by each generator is determined, as is the total electric power produced by the river system as a whole.

The relations between the different variables are not necessarily linear. The volume versus level characteristic of a reservoir is nonlinear and depends on the shape of the reservoir. The power of the turbine depends on the turbine flow and the head, i.e., the height difference between the levels of the upper and lower reservoir. Even for constant head, the relation between turbine flow and power produced is nonlinear. The efficiency of a generator depends nonlinearly on the power produced by the turbine. The flow into and out of a reservoir involves certain delays (the time that water entering the reservoir takes until it is available at the turbines), which in turn depend on the water level in the reservoir. In river segments between reservoirs, the velocity of the water flow depends on its volume. Each machine requires a certain minimum power for operation; that is, a machine can either be operated with at least a certain level of power, or it has to be shut down and not operated at all. The velocity of load changes in turbines may also have to be restricted. The latter operational constraints are nonconvex.

The nonlinear and nonconvex relations of practical problems can be tackled by using mixed-integer formulations that represent nonlinearities by piecewise linear functions with a sufficient number of linear segments.

Let $x_i(t), i = 1, \ldots, k$, be the reservoir volumes of the k reservoirs of the hydro system; $\bar{u}_i(t), i = 1, \ldots, k$, the spill flows from the reservoirs i; $u_i(t), i = 1, \ldots, k$, be the turbine flows through the turbines of reservoir i; and $\tilde{q}_i(t)$ be the exogenous inflows into each reservoir of the hydro system. Let $y_i(t), i = 1, \ldots, k$, be the electric power produced at each reservoir and $y(t)$ be the total electric power produced by the system at time t. The optimal control problem of the hydro power system will be formulated in terms of these variables. Let $c(t)$ be the electricity price that can be achieved at the market.

In order not to blur the insight into the problem with too many details, we make a few simplifications in the formulation presented below. As a river system, we consider a series of reservoirs. We consider constant flow times τ_i; we do not take into account minimum load requirements for the operation of generator-turbine units; and we formulate the functions $f_{i,}$ which relate the electricity generated to the turbine flows and the head of the reservoir, by a piecewise linear representation. Under these simplifications, the hydro system can be represented by a system of linear differential equations, and the problem appears as a linear optimal control problem with a fixed endpoint:

$$\max \int_0^T c(t)y(t)\, dt$$

subject to
$\quad i = 1, \ldots, k :$

$$\dot{x}_i(t) = \tilde{q}_i(t) - \bar{u}_i(t) - u_i(t) + \bar{u}_{i-1}(t - \tau_i) + u_{i-1}(t - \tau_i),\ u_0(t) = \bar{u}_0(t) = 0,$$

$$x_i(0) = x_i^0, x_i(T) = x_i^T,\ \text{given:}$$

$$u_i^{\min} \le u_i(t) \le u_i^{\max}, \bar{u}_i^{\min} \le \bar{u}_i(t),$$

$$y_i(t) = f_i(x_i, x_{i+1}, u_i),$$

$$y(t) = \sum_{i=1}^k y_i(t),$$

$$y^{\min}(t) \le y(t) \le y^{\max}(t).$$

Note that the inflows $\tilde{q}_i(t)$ are uncertain parameters. Initially, only a forecast of the inflows is available. After some time, observations of the random parameters are made. As the control system is supposed to be able to learn, it has to be able to recognize differences between the expected values and later observed outcomes of inflows and adjust the control decisions, $u_i(t)$ and $\bar{u}_i(t)$.

Instead of solving the problem directly as a continuous time problem, we discretize the planning horizon into T discrete time steps. The choice of time-step width depends on the particular model to be solved and its parameters. For example, in our application to hydro power systems control, we will probably choose a time-step width of about 10 minutes. The planning horizon of 3 hours then breaks down into 18 time steps.

By discretizing, we can transform the system of linear differential equations into a system of linear equations, and the optimal control problem can be stated as a multi-stage dynamic linear program. The random right-hand sides correspond to uncertain exogenous inflows. When the control problem is stated and solved, only the predicted values of inflows are known. Due to prediction error the forecasted values of inflows are random parameters. We can assume that their distribution is known.

In general, we expect dependency between the stochastic parameters, both within a certain stage and between stages. We describe correlation of random parameters within a stage by a linear relation; e.g., outcomes of the k uncertain inflows $\tilde{q}_t$ can be obtained by multiplying outcomes of a vector of independent random parameters $V = (V_1, \ldots, V_h)$ by a matrix $F(k \times h)$ so that

$$\tilde{q}^t = FV^t.$$

We consider interperiod dependency as a Markovian process (additive dependency):

$$\tilde{q}^t = FV^t + H\tilde{q}^{t-1},$$

where $H(k \times k)$. The value of random parameter i in period t is a weighted sum of the values of the random parameters in the previous period, $\tilde{q}^{t-1}$, plus a weighted sum of some independent random variation in the current period, V^t. The parameters of this linear additive correlation model can be estimated based on historical observations.

4.8.2   The Multiperiod Asset Allocation Problem

In Dantzig and Infanger (1991) [28] we formulated a class of multiperiod financial asset allocation problems related to that of Mulvey and Vladimirou (1989) [96] and showed how they can be solved by adaptations of multi-stage stochastic linear programming methodology. We now outline this application.

At time period 1 a certain amount of wealth is available to a decision maker invested in assets $i = 1, \ldots, n$ and in cash, which we index as asset $n + 1$. We denote $\bar{x}_i, i = 1, \ldots, n + 1$, to be the dollar value of the initially available assets. The decision maker has to decide each period how to rearrange his portfolio so as to achieve the best return on his initial investment over time. We consider the problem in discrete time and define time steps $t = 1, \ldots, T$, with T being the end of the planning horizon.

At each time period t, the investor can either maintain the level of asset i, buy more, or sell off part (or all) of it. We denote by y_i^t the amount sold of asset i in period t and by x_i^t the amount of asset i retained in period t. Selling asset i means decreasing the value of x_i^t and increasing the value of cash, x_{n+1}^t. Also, the investor has the choice of using available cash to buy certain amounts of assets i. The amount bought in period t is denoted by z_i^t.

Buying and selling entails transaction costs, which we assume to be proportional to the dollar value of the trade. We denote by $100\nu_i$ the transaction costs (expressed as a percentage) associated with buying one unit of asset i and by $100\mu_i$ the transaction costs (expressed as a percentage) associated with selling one unit of asset i. Buying one unit of asset i requires $1 + \nu_i$ units of cash, and selling one unit of asset i returns $1 - \mu_i$ units of cash.

Through buying and selling the investor can restructure his portfolio in each time period t. Once this stage t decision is made, the post-trade holdings x_i^t, $i = 1, \ldots, n+1$, can be calculated. The shares in the portfolio are then kept constant until the next time period. The value of x_i^t is affected by the returns on the market. For example, a holding of x_i^t at time t changes in value to $R_i^t x_i^t$, where R_i^t denotes the return factor from period t to period $t + 1$.

At time t, when the decision on rearranging the portfolio has to be made, returns R_i^t for $i = 1, \ldots, n$, are not known to the decision maker with certainty. Only the return on cash, R_{n+1}^t, is assumed known. However, we assume we know the probability distributions of the R_i^t. The problem is of the "wait and see" type. The decision in period t has to be made on the basis of distributions of future returns R_i^s, for $i = 1, \ldots, n$, $s = t, \ldots, T$, but the values of prior returns R_i^s, $i = 1, \ldots, n$, $s = 1, \ldots, t-1$, have already been observed. We denote by $R^t = R_i^t$, for $i = 1, \ldots, n$, the n-dimensional random vector of returns, with outcomes $r^t(\omega_t)$, $\omega_t \in \Omega_t$, and corresponding probabilities p^{ω_t}. Ω_t is the set of all possible outcomes in t. The random returns R_i^t of period t are mutually dependent and also dependent on the random returns in the previous period.

After the last period, T, no decision is made. Only the value of the portfolio is determined by adding the values of all assets, including the last-period returns. We call this value v^T. The goal of the decision maker is to maximize $E u(v^T)$, the expected utility of the value of the portfolio at the end of period T. The utility function $u(v^T)$ captures the way the investor views risk. If $u(v^T)$ is linear, it reflects risk neutrality; if $u(v^T)$ is concave, it models risk averseness. Nonlinear utility functions require nonlinear programming techniques for the solution of the problem. Our methodology is not restricted to linear problems. However, for the sake of ease and computational speed, we approximate the nonlinear function by a piecewise linear function with a sufficiently large number of linear segments.

In the model presented here we do not consider short selling of assets, although this feature could be easily incorporated. We also do not consider borrowing of cash, which also could be easily incorporated. The holdings of assets, as well as the amounts of assets sold or bought, have to be positive. In general there are also lower ($\underline{x}$) and upper ($\overline{x}$) bounds on holdings, as well as on amounts of assets to be sold ($\underline{y}, \overline{y}$) or to be bought ($\underline{z}, \overline{z}$), which are given by the investor and/or by the market. For example, a certain asset may be available only up to a certain amount, or an investor may want to have a certain asset constitute at least a certain amount of dollar value in the portfolio. Therefore in general we formulate $\underline{x}_i^t \le x_i^t \le \overline{x}_i^t$, $\underline{y}_i^t \le y_i^t \le \overline{y}_i^t$, $\underline{z}_i^t \le z_i^t \le \overline{z}_i^t$, where $\underline{x}_i^t \ge 0$, $\underline{y}_i^t \ge 0$, $\underline{z}_i^t \ge 0$, $r_i^0 x_i^0$ given for $i = 1, \ldots, n+1$, $t = 1, \ldots, T$.

We can now state the model:

$t = 1, \ldots, T, \ i = 1, \ldots, n+1, \ r_i^0 x_i^0$ given:

$$-r_i^{t-1} x_i^{t-1} \ + \ x_i^t \ + \ y_i^t \ - \ z_i^t \ = \ 0, \ i = 1, \ldots, n$$

$$-r_{n+1}^{t-1} x_{n+1}^{t-1} \ + \ x_{n+1}^t \ - \ \sum_{i=1}^n (1 - \mu_i) y_i^t \ + \ \sum_{i=1}^n (1 + \nu_i) z_i^t \ = \ 0,$$

$$-\sum_{i=1}^{n+1} r_i^T x_i^T \ + \ v^T \ = \ 0,$$

$$\max E \, u(v^T)$$

$$\underline{x}_i^t \leq x_i^t \leq \overline{x}_i^t, \quad \underline{y}_i^t \leq y_i^t \leq \overline{y}_i^t, \quad \underline{z}_i^t \leq z_i^t \leq \overline{z}_i^t, \quad i = 1, \ldots, n, \quad t = 1, \ldots, T.$$

We describe correlation between asset returns using a factor model. Using factors is common in the financial industry (e.g., Perold (1984) [108]); hence, historical data on various factors are commercially available. The idea of the factor model is to relate the vector of asset returns $R^t = (R_1, \ldots, R_n)^t$ to factors $V^t = (V_1, \ldots, V_h)^t$. Although the number of assets, n, is large (a model should be able to handle about 500 to 3000 assets), the number of factors, h, is comparatively small. Factor models used in the financial industry typically involve no more than 20 different time series called factors. The $n \times h$ factor matrix F relates R^t to V^t:

$$R^t = FV^t.$$

The coefficients of the factor matrix are estimated using regression analyses on historical data. By linear transformations of historical factors, the transformed factors can always be determined in such a way that the factors V^t are orthogonal. These factors can then be interpreted as **independent** random parameters assumed to be normally distributed or log-normally distributed. Using the factor model, we can generate stochastically dependent returns on the computer by using these stochastically independent factors. We denote an outcome of the random factor V_i^t by v_i^t, with corresponding probability $p(v_i^t) := \text{prob } (V_i^t = v_i^t)$. Up to three stages, we can also formulate inter-stage dependency of the returns, using the auto-correlative dependency model.

As one can now see easily, the multiperiod asset model proposed fits exactly into the framework of a general class of multi-stage stochastic linear programs with recourse. The factor model for generating dependent returns and the auto-correlative inter-stage dependency model define special classes of dependencies between stochastic parameters.

Conclusion

We have discussed and developed a novel approach for solving large-scale stochastic linear programs based on a combination of dual (Benders) decomposition and Monte Carlo importance sampling. Numerical results from large-scale test problems in the areas of facility expansion planning and financial planning demonstrated that very accurate solutions of stochastic linear programs can be obtained with only a small sample size.

The large-scale test problems included various stochastic parameters. For example, the largest problem representing expansion planning for multiarea electric power systems included 39 stochastic parameters. In the deterministic equivalent formulation, if it were possible to state it, the problem would appear as a linear program with about 4.5 billion constraints and variables. The largest portfolio optimization problem included 52 stochastic parameters, which in the deterministic equivalent formulation would appear as a linear program with about 10^{27} constraints and a similar number of variables. Problems this large hitherto seemed to be intractable. Using our method, however, we have been able to solve them on a laptop 80386 computer.

The test results indicate that we have not yet reached the limits of the approach. The sample sizes turned out to be so small that use of parallel processors is not a *sine qua non* condition for solving even large-scale stochastic linear problems. In order to speed up the computation time in the case where large sample sizes are required, we have developed a parallel implementation running on a hypercube multicomputer. The numerical results show that speedups of about 60% can be obtained using 64 parallel processors.

Encouraged by the promising numerical results for two-stage and a restricted class of multi-stage problems, we have developed the theory for a general class of multi-stage stochastic linear programs. Our approach for solving multi-stage problems includes special sampling techniques for computing upper bounds and methods of sharing cuts between different subproblems. It will enable us to efficiently solve large-scale multi-stage problems with many stages and numerous stochastic parameters in each stage. The implementation is subject to future research, but preliminary numerical results have turned out to be promising.

Further research includes improved decomposition techniques for large-scale problems (e.g., optimized tree-traversing strategies and passing information based on nonoptimal subproblems), improvements to the importance sampling approach (e.g., using different types of approximation functions), improved software (e.g., a parallel implementation of the multi-stage algorithm on distributed workstations), and the testing of the methodology on different practical problems in different areas.

Bibliography

[1] Abrahamson, P.G. (1983): A Nested Decomposition Approach for Solving Staircase Linear Programs, Technical Report SOL 83-4, Department of Operations Research, Stanford University, Stanford, CA.

[2] Ariyawansa, K.A., and Hudson, D.D. (1990): Performance of a Benchmark Parallel Implementation of the Van Slyke and Wets Algorithm for Two-Stage Stochastic Programs on the Sequent/Balance, *Concurrency: Practice and Experience 3* (2), 109–128.

[3] Avriel, M., Dantzig, G.B., and Glynn, P.W. (1989): Decomposition and Parallel Processing Techniques for Large-Scale Electric Power System Planning under Uncertainty, *Proceedings of the Workshop on Resource Planning under Uncertainty*, Stanford University, Stanford, CA.

[4] Beale, E.M.L. (1955): On Minimizing a Convex Function Subject to Linear Inequalities, *Journal of the Royal Statistical Society, Series B 17*, 173–184.

[5] Beale, E.M.L., Dantzig, G.B., and Watson R.D. (1986): A First-Order Approach to a Class of Multi-Time-Period Stochastic Programming Problems, *Mathematical Programming Study 27*, 103–117.

[6] Benders, J.F. (1962): Partitioning Procedures for Solving Mixed-Variable Programming Problems, *Numerische Mathematik 4*, 238–252.

[7] Ben-Tal, A., and Hochman, E. (1972): More Bounds on the Expectation of a Random Variable, *Journal of Applied Probability 9*, 803–812.

[8] Birge, J.R. (1980): Solution Methods for Stochastic Dynamic Linear Programs, Technical Report SOL 80-29, Department of Operations Research, Stanford University, Stanford, CA.

[9] Birge, J.R. (1984): Aggregation in Stochastic Linear Programming, *Mathematical Programming 31*, 25–41.

[10] Birge, J.R. (1985): Decomposition and Partitioning Methods for Multi-Stage Stochastic Linear Programming, *Operations Research 33*, 989–1007.

131

[11] Birge, J.R., and Louveaux, F.V. (1985): A Multicut Algorithm for Two-Stage Linear Programs, Technical Report, Department of IOE, University of Michigan, Ann Arbor, MI.

[12] Birge, J.R., and Teboulle, M. (1989): Upper Bounds on the Expected Values of a Convex Function Using Gradient and Conjugate Function Information, *Mathematics of Operations Research 14/4*, 745–759.

[13] Birge, J.R., and Wallace, S.W. (1988): A Separable Piecewise Linear Upper Bound for Stochastic Linear Programs, *SIAM Journal on Control and Optimization 26*, 3.

[14] Birge, J.R., and Wets, R.J. (1986): Designing Approximation Schemes for Stochastic Optimization Problems, in Particular for Stochastic Programs with Recourse, *Mathematical Programming Study 27*, 54–102.

[15] Birge, J.R., and Wets, R.J. (1987): Computing Bounds for Stochastic Programming Problems by Means of a Generalized Moment Problem, *Mathematics of Operations Research 12*, 149–162.

[16] Birge, J.R., and Wets, R.J. (1989): Sublinear Upper Bounds for Stochastic Programs with Recourse, *Mathematical Programming 43*, 131–149.

[17] Birge, J.R., and Wets, R.J. (eds.) (1991): Stochastic Programming, Proceedings of the 5th International Conference on Stochastic Programming, *Annals of Operations Research 30 and 31*.

[18] Charnes, A., and Cooper, A.A. (1959): Chance Constrained Programming, *Management Science 6*, 73–79.

[19] Cipra, T. (1985): Moment Problem with Given Covariance Structure in Stochastic Programming, *Československá Akademie Věd. Ekonomicko-Matematicky Obzor 21*, 66–77.

[20] Dantzig, G.B. (1948): Programming in a Linear Structure, Comptroller, USAF, Washington, D.C.

[21] Dantzig, G.B. (1955): Linear Programming under Uncertainty, *Management Science 1*, 197–206.

[22] Dantzig, G.B. (1963): *Linear Programming and Extensions*, Princeton University Press, Princeton, NJ.

[23] Dantzig, G.B. (1988): Planning under Uncertainty Using Parallel Computing, *Annals of Operations Research 14*, 1–16.

[24] Dantzig, G.B., and Glynn, P.W. (1990): Parallel Processors for Planning under Uncertainty, *Annals of Operations Research 22*, 1–21.

[25] Dantzig, G.B., Glynn, P.W., Avriel, M., Stone, J., Entriken, R., and Nakayama, M. (1989): Decomposition Techniques for Multi-Area Generation and Transmission Plan-

ning under Uncertainty, EPRI Report 2940-1, Electric Power Research Institute, Palo Alto, CA.

[26] Dantzig, G.B., Ho, J.K., and Infanger, G. (1991): Solving Stochastic Linear Programs on a Hypercube Multicomputer, Technical Report SOL 91-10, Department of Operations Research, Stanford University, Stanford, CA.

[27] Dantzig, G.B., and Infanger, G. (1991): Multi-Stage Stochastic Linear Programs for Portfolio Optimization, Technical Report SOL 91-11, Department of Operations Research, Stanford University, Stanford, CA; to appear in *Annals of Operations Research*.

[28] Dantzig, G.B., and Infanger, G.(1991): Large-Scale Stochastic Linear Programs: Importance Sampling and Benders Decomposition, Technical Report SOL 91-4, Department of Operations Research, Stanford University, Stanford, CA.

[29] Dantzig, G.B., and Madansky, M. (1961): On the Solution of Two-Staged Linear Programs under Uncertainty, *Proceedings of the 4th Berkeley Symposium on Mathematical Statistics and Probability I*, ed. J. Neyman, 165–176.

[30] Dantzig, G.B., and Pereira, M.V.F. (1988): Mathematical Decomposition Techniques for Power System Expansion Planning, EPRI Report EL-5299, Electric Power Research Institute, Palo Alto, CA.

[31] Dantzig, G.B., and Wolfe, P. (1960): The Decomposition Principle for Linear Programs, *Operations Research 8*, 110–111.

[32] Davis, P.J., and Rabinowitz, P. (1984): Methods of Numerical Integration, Academic Press, London.

[33] Deák, I. (1988): Multidimensional Integration and Stochastic Programming, in Ermoliev, Y., and R.J.-B. Wets (eds.): *Numerical Techniques for Stochastic Optimization*, Springer Verlag, Berlin, 187–200.

[34] Dempster, M.A.H. (1980): Introduction to Stochastic Programming, in Dempster, M.A.H. (ed.): *Stochastic Programming*, Academic Press, 3–59.

[35] Dempster, M.A.H. (1986): On Stochastic Programming II: Dynamic Problems under Risk, Research Report DAL TR 86-5, Dalhousie University, Canada.

[36] Dupačowá, J. (1978): Minimax Approach to Stochastic Linear Programming and the Moment Problem, *Zeitschrift für Angewandte Mathematik und Mechanik 58T*, 466–467.

[37] Dupačowá, J., and Wets, R.J.-B. (1988): Asymptotic Behavior of Statistical Estimators of Optimal Solutions of Stochastic Optimization Problems, *Annals of Mathematical Statistics 16*, 1517–1549.

[38] Edmundson, H.P. (1956): Bounds on the Expectation of a Convex Function of a Random Variable, Paper 982, Rand Corporation, Santa Monica, CA.

[39] Entriken, R. (1988): A Parallel Decomposition Algorithm for Staircase Linear Programs, Report ORNL/TM 11011, Oak Ridge National Laboratory, Oak Ridge, TN.

[40] Entriken, R. (1989): The Parallel Decomposition of Linear Programs, Technical Report SOL 89-17, Department of Operations Research, Stanford University, Stanford, CA.

[41] Entriken, R., and Infanger, G. (1990): Decomposition and Importance Sampling for Stochastic Linear Models, *Energy, The International Journal*, Vol. 15, No. 7/8, 645–659.

[42] Ermoliev, Y. (1983): Stochastic Quasi-Gradient Methods and Their Applications to Systems Optimization, *Stochastics 9*, 1–36.

[43] Ermoliev, Y. (1988): Stochastic Quasi-Gradient Methods, in Ermoliev, Y., and Wets, R.J.-B. (eds.): *Numerical Techniques for Stochastic Optimization*, Springer Verlag, Berlin, 141–186.

[44] Ermoliev, Y. (1988): Stochastic Quasigradient Methods and Their Implementation, in Ermoliev, Y., and Wets, R.J.-B. (eds.): *Numerical Techniques for Stochastic Optimization*, Springer Verlag, Berlin, 313–351.

[45] Ermoliev, Y., Gaivoronski, A., and Nedeva, C. (1985): Stochastic Optimization Problems with Partially Known Distribution Functions, *SIAM Journal on Control and Optimization 23*, 696–716.

[46] Ermoliev, Y., and Wets, R.J.-B. (eds.) (1988): *Numerical Techniques for Stochastic Optimization*, Springer Verlag, Berlin.

[47] Ferguson, A., and Dantzig, G.B. (1956): The Allocation of Aircraft to Routes: An Example of Linear Programming under Uncertain Demand, *Management Science 3*, 45–73.

[48] Frauendorfer, K. (1988): Solving SLP Recourse Problems with Arbitrary Multivariate Distributions—The Dependent Case, *Mathematics of Operations Research 13*, No. 3, 377–394.

[49] Frauendorfer, K. (1992): *Stochastic Two-Stage Programming*, Lecture Notes in Economics and Mathematical Systems 392, Springer Verlag, Berlin.

[50] Frauendorfer, K., and Kall, P. (1988): Solving SLP Recourse Problems with Arbitrary Multivariate Distributions—The Independent Case, *Problems of Control and Information Theory*, Vol. 17 (4), 177–205.

[51] Gaivoronski, A. (1988): Implementation of Stochastic Quasigradient Methods, in Ermoliev, Y., and Wets, R.J.-B. (eds.): *Numerical Techniques for Stochastic Optimization*, Springer Verlag, Berlin.

[52] Gaivoronski, A., and Nazareth, J.L. (1989): Combining Generalized Programming and Sampling Techniques for Stochastic Programs with Recourse, in Proceedings of the Work-

VIA: UP

PRIME-INDUCT
-SLSB

LOCATION	QTY	ISBN	AUTHOR/TITLE
O-11C-001-42	1	0-89426-249-1	INFANGER PLAN UNDER UNCERTAINTY

WAREHOUSE INSTRUCTIONS

SLA: 7   BOX: Staple

those words
Rise up. When they turn & leave the page
With trails of colors too
Heat or cold, fear, longing
Parting, sunrise, faint fragrance
Of love regained
From one page,
Well read

shop on Resource Planning under Uncertainty for Electric Power Systems, Department of Operations Research, Stanford University, Stanford, CA.

[53] Gassmann, H. (1990): MSLiP: A Computer Code for the Multi-Stage Stochastic Linear Programming Problem, *Mathematical Programming 47*, 407–423.

[54] Gassmann, H., and Ziemba, W.T. (1986): A Tight Upper Bound for the Expectation of a Convex Function of a Multivariate Random Variable, *Mathematical Programming Study 27*, 39–52.

[55] Geoffrion, A.M. (1974): Elements of Large-Scale Mathematical Programming, *Management Science 16*, No. 11.

[56] Gill, P.E., Murray, W., and Wright, M.H. (1981): *Practical Optimization*, Academic Press, London.

[57] Gill, P.E., Murray, W., and Wright, M.H. (1991): *Numerical Linear Algebra and Optimization*, Addison-Wesley, Redwood City, CA.

[58] Glynn, P.W., and Iglehart, D.L. (1989): Importance Sampling for Stochastics Simulation, *Management Science 35*, 1367–1392.

[59] Hall, P. (1982): Rates of Convergence in the Central Limit Theorem, *Research Notes in Mathematics 62*.

[60] Hammersly, J.M., and Handscomb, D.C. (1964): *Monte Carlo Methods*, Mathuen, London.

[61] Higle, J.L., and Sen, S. (1989): Stochastic Decomposition: An Algorithm for Two-Stage Linear Programs with Recourse, *Mathematics of Operations Research 16/3*, 650–669.

[62] Higle, J.L., and Sen, S. (1989): Statistical Verification of Optimality Conditions for Stochastic Programs with Recourse, *Annals of Operations Research 30*, 215–240.

[63] Higle, J.L., Sen, S., and Yakowitz, D. (1990): Finite Master Programs in Stochastic Decomposition, Technical Report, Dept. of Systems and Industrial Engineering, The University of Arizona, Tucson, AZ.

[64] Higle, J.L., Lowe, W., and Odio, R. (1990): Conditional Stochastic Decomposition: An Algorithmic Interface for Optimization/Simulation, Working Paper 90-009, Dept. of Systems and Industrial Engineering, The University of Arizona, Tucson, AZ.

[65] Glassey, R. (1973): Nested Decomposition and Multi-Stage Linear Programs, *Management Science 20*, 282–292.

[66] Hiller, R.S., and Eckstein, J. (1990): Stochastic Dedication: Designing Fixed Income Portfolios Using Massively Parallel Benders Decomposition, Working Paper 91-025, Harvard Business School, Cambridge, MA.

[67] Ho, J.K., and Gnanendran, S.K. (1989): Distributed Decomposition of Block-Angular Linear Programs on a Hypercube Computer, College of Business Administration, University of Tennessee, Knoxville, TN.

[68] Ho, J.K., Lee, T.C., and Sundarraj, R.P. (1988): Decomposition of Linear Programs Using Parallel Computation, *Mathematical Programming 42*, 391–405.

[69] Ho, J.K., and Loute, E. (1981): A Set of Staircase Linear Programming Test Problems, *Mathematical Programming 20*, 245–250.

[70] Ho, J.K., and Manne, A.S. (1974): Nested Decomposition for Dynamic Models, *Mathematical Programming 6*, 121–140.

[71] Huang, C.C., Ziemba, W.T., and Ben-Tal, A. (1977): Bounds on the Expectation of a Convex Function with a Random Variable with Applications to Stochastic Programming, *Operations Research 25*, 315–325.

[72] IBM (1991): *Optimization Subroutine Library, Guide and Reference, Release 2*, No. SC23-0519-02, IBM Corporation, Kingston, NY.

[73] Infanger, G. (1990): Monte Carlo (Importance) Sampling within a Benders Decomposition Algorithm for Stochastic Linear Programs, Technical Report SOL 89-13R, Department of Operations Research, Stanford University, Stanford, CA.

[74] Infanger, G. (1991): Monte Carlo (Importance) Sampling within a Benders Decomposition Algorithm for Stochastic Linear Programs, *Annals of Operations Research, 39*.

[75] Intel Corporation (1988): *iPSC/2 Fortran Programmer's Reference Manual*, Order No. 311019-003.

[76] Intel Corporation (1988) *iPSC/2 Green Hills Fortran Language Reference Manual* (Preliminary), Order No. 311020-003.

[77] Jensen, J.L. (1906): Sur les fonctions convexes et les inegalites entres les valeurs moyennes, *Acta Mathematica 30*, 175–193.

[78] Kall, P. (1974): Approximations to Stochastic Programs with Complete Fixed Recourse, *Numerische Mathematik 22*, 333–339.

[79] Kall, P. (1976): *Stochastic Linear Programming*, Springer Verlag, Berlin.

[80] Kall, P. (1979): Computational Methods for Two-Stage Stochastic Linear Programming Problems, *Zeitschrift für Angewandte Mathematik und Physik 30*, 261–271.

[81] Kall, P., Ruszczynski, A., and Frauendorfer, K. (1988): Approximation Techniques in Stochastic Programming, in Ermoliev, Y., and Wets, R.J.-B. (eds.): *Numerical Techniques for Stochastic Optimization*, Springer Verlag, Berlin.

[82] Kall, P., and Stoyan, D. (1982): Solving Stochastic Programming Problems with Recourse Including Error Bounds, *Mathematische Operationsforschung und Statistik, Series Optimization, 13*, 431–447.

[83] King, A., and Wets, R.J.-B. (1989): EPI-Consistency of Convex Stochastic Programs, submitted to *Stochastics*.

[84] Krishna, A.S. (1993): Importance Sampling Techniques for Stochastic Linear Programming, Dissertation, Technical Report SOL 93-00, Department of Operations Research, Stanford University, Stanford, CA.

[85] Kusy, M.I., and Ziemba, W.T. (1986): A Bank Asset and Liability Management Model, *Operations Research 34*, 356–376.

[86] Lasdon, L. (1970): *Optimization Theory for Large Systems*, Macmillan, New York.

[87] Lavenberg, S.S., and Welch, P.D. (1981): A Perspective on the Use of Control Variables to Increase the Efficiency of Monte Carlo Simulation, *Management Science 27*, 322–335.

[88] Louveaux, F.V. (1986): Multi-Stage Stochastic Programs with Block-Separable Recourse, *Mathematical Programming Study 28*, 48–62.

[89] Louveaux, F.V., and Smeers, Y. (1988): Optimal Investment for Electricity Generation: A Stochastic Model and a Test Problem, in Ermoliev, Y., and Wets, R.J.-B. (eds.): *Numerical Techniques for Stochastic Optimization*, Springer Verlag, Berlin.

[90] Lustig, I.J, Mulvey, J.M., and Carpenter, T.J. (1991): Formulating Two-Stage Stochastic Programs for Interior Point Methods, *Operations Research 39*, 757–770.

[91] Madansky, A. (1959): Bounds on the Expectation of a Convex Function of a Multivariate Random Variable, *Annals of Mathematical Statistics 30*, 743–746.

[92] Markowitz, H. (1959): Portfolio Selection: Efficient Diversification of Investments, John Wiley and Sons, New York.

[93] Marti, K. (1980): Solving Stochastic Linear Programs by Semi-Stochastic Approximation Algorithms, in Kall, P., and Prekopa, A. (eds.): *Recent Results in Stochastic Programming*, Lecture Notes in Economics and Mathematical Systems 179, Springer Verlag, Berlin, 191–213.

[94] Morton, D.P. (1993): Algorithmic Advances in Multi-stage Stochastic Programming, Dissertation, Technical Report 93-6, Department of Operations Research, Stanford University, Stanford, CA.

[95] Mulvey, J.M. (1987): Nonlinear Network Models in Finance, *Advances in Mathematical Programming and Financial Planning, 1*, 253.

[96] Mulvey, J.M., and Vladimirou, H. (1989): Stochastic Network Optimization Models for Investment Planning, *Annals of Operations Research 20 (1989)*, 187–217.

[97] Mulvey, J.M., and Vladimirou, H. (1991): Applying the Progressive Hedging Algorithm to Stochastic Generalized Networks, *Annals of Operations Research 31*, 399–424.

[98] Mulvey, J.M., and Vladimirou, H. (1991): Solving Multi-Stage Stochastic Networks: An Application of Scenario Aggregation, *Networks, 21*, 619–643.

[99] Murtagh, B.A., and Saunders, M.A. (1983): MINOS 5.0 User's Guide, SOL 83-20, Department of Operations Research, Stanford University, Stanford, CA.

[100] Nazareth, J.L., and Wets, R.J.-B. (1986): Algorithms for Stochastic Programs: The Case of Nonstochastic Tenders, *Mathematical Programming Study 28*, 48–62.

[101] Nazareth, J.L., and Wets, R.J.-B. (1988): Nonlinear Programming Techniques Applied to Stochastic Programs with Recourse, in Ermoliev, Y., and Wets, R.J.-B. (eds.): *Numerical Techniques for Stochastic Optimization*, Springer Verlag, Berlin, 95–122.

[102] Niederreiter, H. (1986): Multidimensional Numerical Intergration Using Pseudo Random Numbers, *Mathematical Programming Study 27*, 17–38.

[103] Pereira, M.V., and Pinto, L.M.V.G. (1989): Stochastic Optimization of a Multi-Reservoir Hydro-Electric System—A Decomposition Approach, *Water Resources Research 21*, 779–792.

[104] Pereira, M.V., Pinto, L.M.V.G., Oliveira, G.C., and Cunha, S.H.F. (1989): A Technique for Solving LP Problems with Stochastic Right-Hand Sides, CEPEL, Centro del Pesquisas de Energia Electria, Rio de Janeiro, Brazil.

[105] Pereira, M.V., Pinto, L.M.V.G., Oliveira, G.C., and Cunha, S.H.F. (1989): A Decomposition Approach to Automated Generation/Transmission Expansion Planning, CEPEL, Centro del Pesquisas de Energia Electria, Rio de Janeiro, Brazil.

[106] Pereira, M.V., and Pinto, L.M.V.G. (1989): Stochastic Dual Dynamic Programming, Technical Note, DEE-PUC/RJ, Catholic University of Rio de Janeiro, Rio de Janeiro, Brazil.

[107] Pereira, M.V.F., and L.M.V.G. Pinto (1991): Multi-Stage Stochastic Optimization Applied to Energy Planning, *Mathematical Programming 52*, 359–375.

[108] Perold, A. (1984): Large-Scale Portfolio Optimization, *Management Science 30*, 10, 1143.

[109] Prékopa, A. (1988): Numerical Solution of Probabilistic Constrained Programming Models, in Ermoliev, Y., and Wets, R.J.-B. (eds.): *Numerical Techniques for Stochastic Optimization*, Springer Verlag, Berlin, 123–139.

[110] Prékopa, A. (1988): Boole-Bonferroni Inequalities and Linear Programming, *Operations Research 36*, 145–162.

[111] Prékopa, A. (1989): Sharp Bounds on Probabilities Using Linear Programming, *Operations Research 38*, 227–239.

[112] Prékopa, A. (1990): The Discrete Moment Problem and Linear Programming, *Discrete Applied Mathematic 27*, 235–254.

[113] Pflug, G.C. (1988): Stepsize Rules, Stopping Times, and Their Implementations in Stochastic Quasigradient Algorithms, in Ermoliev, Y., and Wets, R.J.-B. (eds.): *Numerical Techniques for Stochastic Optimization*, Springer Verlag, Berlin, 353–372.

[114] Robinson, S.M., and Wets, R.J.-B. (1987): Stability in Two-Stage Stochastic Programming, *SIAM Journal on Control and Optimization 25*, 1409–1416.

[115] Rockafellar, R.T., and Wets, R.J.-B. (1989): Scenario and Policy Aggregation in Optimization under Uncertainty, *Mathematics of Operations Research 16*, 119–147.

[116] Rubinstein, R.Y., and Marcus, R. (1985): Efficiency of Multivariate Control Variates in Monte Carlo Simulation, *Operations Research 33*, 661–677.

[117] Ruszczynski, A. (1986): A Regularized Decomposition Method for Minimizing a Sum of Polyhedral Functions, *Mathematical Programming 35*, 309–333.

[118] Ruszczynski, A. (1987): A Linearization Method for Nonsmooth Stochastic Programming Problems, *Mathematics of Operations Research 12*, 32–49.

[119] Strazicky, B. (1980): Computational Experience with an Algorithm for Discrete Recourse Problems, in Dempster, M. (ed.): *Stochastic Programming*, Academic Press, London, 263–274.

[120] Tomlin, J. (1973): LPM1 User's Guide, Manuscript, Systems Optimization Laboratory, Stanford University, Stanford, CA.

[121] Van Slyke, R.M., and Wets, R.J.-B. (1966): Programming under Uncertainty and Stochastic Optimal Control, *SIAM Journal on Control and Optimization 4*, 179–193.

[122] Van Slyke, R.M., and Wets, R.J.-B. (1969): L-Shaped Linear Programs with Applications to Optimal Control and Stochastic Programming, *SIAM Journal on Applied Mathematics 17*, 638–663.

[123] Wallace, S.W. (1987): A Piecewise Linear Upper Bound on the Network Recourse Problem, *Mathematical Programming 38*, 133–146.

[124] Wets, R.J.-B. (1966): Programming under Uncertainty: The Complete Problem, *Zeitschrift für Wahrscheinlichkeit und verwandschaftliche Gebiete 4*, 316–339.

[125] Wets, R.J.-B. (1974): Stochastic Programs with Fixed Recourse: The Equivalent Deterministic Program, *SIAM Review 16*, 309–339.

[126] Wets, R.J.-B. (1983): Solving Stochastic Programs with Simple Recourse, *Stochastics 10*, 219–242.

[127] Wets, R.J.-B. (1985): On Parallel Processors Design for Solving Stochastic Programs, *Proceedings of the 6th Mathematical Programming Symposium*, Japanese Mathematical Programming Society, Japan, 13–36.

[128] Wets, R.J.-B. (1988): Large-Scale Linear Programming Techniques in Stochastic Programming, in Ermoliev, Y., and Wets, R.J.-B. (eds.): *Numerical Techniques for Stochastic Optimization*, Springer Verlag, Berlin.

[129] Wets, R.J.-B. (1989): Stochastic Programming, in Nemhauser, G.L., Kan, A.H.G., and Todd, M.J. (eds.): *Handbook on Operations Research and Management Science*, North-Holland, Amsterdam, 573–629.

[130] Wittrock, R.J. (1983): Advances in a Nested Decomposition Algorithm for Solving Staircase Linear Programs, Technical Report SOL 83-2, Department of Operations Research, Stanford University, Stanford, CA.

[131] Zenios, S.A. (1990): Massively Parallel Algorithms for Financial Modelling under Uncertainty, Decision Sciences Department, The Wharton School, University of Pennsylvania, Philadelphia, PA.

[132] Zenios, S.A. (1992): A Model for Portfolio Management with Mortgage-Backed Securities, Decision Sciences Department, The Wharton School, University of Pennsylvania, Philadelphia, PA; to appear in *Annals of Operations Research*.